A FUTURE SHAPED BY AI AND ROBOTS

Exploring the Promises, Perils,
and Ethical Dilemmas of Emerging Technologies

By Ihab Hassan

COPYRIGHT

Copyright © 2024 by Ihab Hassan

All rights reserved. No part of this publication may be reproduced, distributed, or transmitted in any form or by any means, including photocopying, recording, or other electronic or mechanical methods, without the prior written permission of the publisher, except in the case of brief quotations in reviews, articles, or other non-commercial uses as permitted by copyright law.

This book is a work of nonfiction. While every effort has been made to ensure accuracy, the author does not assume responsibility for errors, omissions, or differing interpretations.

First Edition: 2024

For permissions and inquiries, contact:
https://www.amazon.com/author/scammed-victims

Published in the United States of America

DEDICATION

To the thinkers, innovators, and dreamers who dare to envision a world where technology transforms lives for the better. And to those who question the ethical boundaries, ensuring that progress always walks hand in hand with responsibility.

ACKNOWLEDGMENTS

This book would not have been possible without the support and encouragement of many individuals. First and foremost, I want to express my heartfelt gratitude to my family and friends. Your endless patience, unwavering belief in my work, and constant encouragement provided me with the motivation to delve deeply into the intricacies of this complex subject.

To my colleagues in the engineering and tech fields, thank you for the stimulating discussions and thought-provoking debates that helped shape many of the ideas presented in this book. Your insights were invaluable, and I truly appreciate the collaborative spirit that we share.

I also owe a debt of gratitude to my research assistants, who tirelessly gathered data, conducted analyses, and ensured that the information presented here is accurate and comprehensive. Your hard work and dedication brought depth and rigor to these pages, and I couldn't have done it without you.

Finally, I would like to extend my sincere thanks to my readers. You are the reason I write. It is my hope that this book not only informs you but also challenges you to think critically about the future we are creating with AI and robotics. Your engagement and curiosity drive the pursuit of knowledge, and I look forward to hearing your thoughts and insights.

INTRODUCTION:

A FUTURE SHAPED BY AI AND ROBOTS

The future promises a world deeply transformed by artificial intelligence (AI) and robotics. Once the stuff of science fiction, these technologies are now an integral part of our reality, with AI systems making critical decisions in industries ranging from finance to healthcare, and robots performing tasks that once required skilled human hands. From autonomous vehicles navigating busy streets to humanoid robots providing companionship and assistance, the future is closer than we might think.

Yet, alongside the exhilarating possibilities of this new era lies a profound sense of uncertainty and, in many cases, fear. What happens when robots malfunction or AI systems make decisions that result in harm rather than benefit? How do we exert control over machines that are potentially smarter than ourselves? These questions become even more pressing when we consider the societal implications of automation, particularly the millions of jobs that may be displaced as machines become capable of performing tasks more efficiently than humans.

This book seeks to explore these critical questions, envisioning a world where robots and AI are as integral to daily life as electricity. We will examine the advancements in technology that bring us to this point, the potential benefits and risks associated with their integration, and the ethical dilemmas that arise when we delegate decision-making to machines.

Moreover, we will confront the darker side of this technological revolution: the fear of losing control over our creations, the ethical challenges posed by autonomous systems, and the catastrophic damage that could occur if these systems fail or are misused. By engaging with these complexities, we aim to foster a nuanced understanding of how AI and robotics are reshaping our lives and what it means for our collective future.

As we embark on this journey, let us challenge ourselves to think critically about the role of technology in our society, considering both its vast potential and its inherent risks. Together, we can navigate this transformative landscape, ensuring that the future we create is one that enhances our humanity rather than diminishes it.

TABLE OF CONTENTS

1. Introduction: A World on the Verge of Change
2. Chapter 1: The Rise of AI in Daily Life
3. Chapter 2: Robotics Revolution — Automation at a Crossroads
4. Chapter 3: How AI is Reshaping Healthcare
5. Chapter 4: AI in Education — Teaching Machines to Teach Us
6. Chapter 5: The Future of Work — Automation, Jobs, and Universal Basic Income
7. Chapter 6: Autonomous Vehicles — Steering the Future of Transportation
8. Chapter 7: The Ethics of AI — Fairness, Bias, and Accountability
9. Chapter 8: Robots and AI in Warfare — The Future of Security and Conflict
10. Chapter 9: The Dangers of AI — How AI Could Harm Society
11. Chapter 10: AI and Human Creativity — Friend or Foe?
12. Chapter 11: AI and Privacy — The Cost of Living in a Data-Driven World
13. Chapter 12: Human-Robot Collaboration — Merging Man and Machine
14. Chapter 13: The Social Divide — AI, Wealth Inequality, and Power Dynamics

15. Chapter 14: Will Robots Replace Us? Sentience and the Singularity Debate
16. Chapter 15: Legal and Regulatory Challenges of AI and Robotics
17. Chapter 16: Environmental Impact of AI and Robotics
18. Chapter 17: Designing Ethical AI — Where Do We Draw the Line?
19. Chapter 18: What Happens When AI Fails? The Risks of Malfunction
20. Chapter 19: AI and the Future of Global Leadership
21. Chapter 20: AI's Role in Social Media and Public Discourse
22. Chapter 21: The Role of AI in Policing and Surveillance
23. Chapter 22: From Factories to Homes — The Invasion of Domestic Robots
24. Chapter 23: The Future of Human Jobs in an AI-Dominated World
25. Chapter 24: AI in Agriculture — Feeding the Future
26. Chapter 25: The AI Arms Race — Global Power Struggles Over Advanced Technologies
27. Chapter 26: AI and the Media — Fake News, Deepfakes, and the War on Truth
28. Chapter 27: AI in Sports — Enhancing Performance or Eroding Integrity?
29. Chapter 28: The Intersection of AI and Human Emotion
30. Chapter 29: The Moral Dilemmas of AI-Driven

Decision Making
31. **Chapter 30: AI and the Reinvention of Art, Music, and Creativity**
32. **Chapter 31: Preparing for a Future with AI — Adapting Society**
33. **Chapter 32: Conclusion: Humanity's Responsibility in Shaping the Future**

CHAPTER 1: THE PROMISE OF AI AND ROBOTICS

In recent decades, **artificial intelligence (AI)** and **robotics** have made a bold leap from the pages of science fiction to the forefront of modern reality. These technologies, once considered distant dreams, are now revolutionizing industries and reshaping our daily lives. From self-driving cars navigating city streets to powerful algorithms predicting everything from stock market trends to human behavior, the promise of AI and robotics is as exhilarating as it is transformative. **But the story is not just one of technological progress it is also a tale of profound societal change and ethical dilemmas.**

The promise of AI and robotics is **thrilling** they can automate processes, solve complex problems, and even offer solutions to some of humanity's most pressing challenges. Yet, these advancements also raise difficult questions: **What will become** of human workers in an AI-driven world? Will machines surpass human intelligence, and if so, what might that mean for the future of human autonomy?** These are no longer theoretical debates but issues we must confront as AI and robotics become increasingly integral to our world.

The Current State of AI and Robotics

At the heart of this technological revolution lies **AI's** unparalleled ability to learn from data, make intelligent decisions, and perform tasks traditionally handled by humans. Through advances in **machine learning**, AI systems can now improve their performance over time, becoming faster, smarter, and more efficient. Robotics, the sibling of AI, involves machines designed to execute specific tasks—often in places that are too dangerous, complex, or tedious for humans. Together, these fields are transforming nearly every aspect of modern society.

AI is everywhere—from the apps that suggest your next purchase to the AI algorithms diagnosing diseases with pinpoint accuracy. **Robotics**, too, is gaining momentum. Industrial robots work tirelessly in factories, while autonomous machines explore remote, dangerous environments like deep-sea oil rigs or disaster zones, places no human can easily reach.

In healthcare, **AI-driven systems** analyze medical data at lightning speed, often identifying diseases such as cancer earlier and more accurately than human doctors. **In agriculture**, autonomous tractors, drones, and irrigation systems are making farming more efficient and less labor-intensive, helping to feed an ever-growing global population. And in **transportation**, companies like Tesla and Waymo are redefining how we think about mobility with autonomous vehicles poised to revolutionize how we travel.

Potential Benefits

The potential benefits of AI and robotics are staggering, promising to not only streamline industries but also radically improve the way we live and work. Here's what's on the horizon:

1. **Unmatched Efficiency**: Automation will revolutionize productivity across countless industries. With AI managing logistics, supply chains will become seamless, optimizing inventory levels and reducing waste. Imagine a world where AI predicts demand for products before consumers even think of buying them.

2. **Smarter Decision-Making**: AI systems can analyze **mountains of data** that would be impossible for humans to process, uncovering patterns and insights that lead to better decisions. In fields like finance and public health, AI is already driving more informed and precise strategies, minimizing risk while maximizing impact.

3. **Cost Reductions on a Global Scale**: Automating repetitive tasks cuts operational costs dramatically, especially in labor-intensive sectors like manufacturing. Robots in factories can work tirelessly without breaks, significantly reducing production costs while enhancing quality and consistency.

4. **A Better Quality of Life**: Robotics is set to **transform everyday living**. From smart homes that manage energy usage and enhance security to assistive robots helping the elderly and disabled live more independently, AI-powered devices will play a central role in making life easier and more enjoyable

for everyone.

5. **Solving Global Challenges**: **AI's potential** to address critical issues like climate change is monumental. From optimizing renewable energy usage to improving agricultural efficiency in drought-prone areas, AI models are crucial for creating a more sustainable, resilient future.

Success Stories: Where AI and Robotics Shine

Across industries, we are already witnessing the transformative power of these technologies:

- **Healthcare**: IBM's Watson Health is at the cutting edge, analyzing vast medical data sets to detect diseases like cancer faster and more accurately than human specialists. Imagine a world where complex diagnoses are delivered in seconds.
- **Transportation**: Autonomous vehicles, spearheaded by Tesla and Waymo, are not just a distant dream—they are becoming a reality. These cars promise to reduce accidents, ease traffic congestion, and reshape urban transportation forever.
- **Agriculture**: With robots tending fields, drones monitoring crops, and AI systems optimizing harvests, farming is becoming more efficient, helping to feed a rapidly expanding global population with fewer resources.
- **Manufacturing**: The rise of **smart factories** is revolutionizing production. Robots, AI-driven processes, and interconnected devices create a seamless, highly efficient manufacturing environment, lowering costs and boosting output.

Conclusion: A Double-Edged Sword

As we stand at the dawn of this technological era, the promise of AI and robotics is undeniably thrilling—but it is also fraught with complexities. Yes, they offer incredible potential to **revolutionize industries**, **boost economies**, and **improve lives**. But with such power comes responsibility. As we embrace these technologies, we must also address the deeper ethical questions: **How will they affect jobs? What rights should AI have? And are we prepared for a future where machines may one day rival human intelligence?**

In the chapters that follow, we will explore these questions and more. The benefits of AI and robotics are profound, but so too are the risks. We will not shy away from the darker side of automation the **job displacement**, the **loss of privacy**, and the ethical dilemmas that come with wielding such powerful tools. Together, we'll navigate the promise and peril of these groundbreaking technologies

CHAPTER 2: THE DARK SIDE OF AI AND ROBOTICS

While the advancements in artificial intelligence (AI) and robotics promise a future of convenience, efficiency, and innovation, they also come with profound challenges and ethical dilemmas. As we stand on the cusp of a technological revolution, it is crucial to examine the potential dangers that accompany these powerful tools. Beyond the awe-inspiring headlines and futuristic dreams lies a reality fraught with risks —one that could reshape society in ways we may not be ready to face.

This chapter delves into the darker side of AI and robotics, focusing on the ethical implications, the potential for misuse, and the societal consequences of unchecked technological advancement.

The Ethical Implications of AI

AI's growing role in decision-making is not without serious ethical concerns. While it enhances our capabilities, it also poses troubling questions about fairness, privacy, and accountability.

1. **Bias and Discrimination**: One of the most alarming issues with AI is its potential to perpetuate and even amplify societal biases. AI systems are only as good as the data they are trained on, and if that data reflects human prejudices, the resulting algorithms can lead to discrimination. For example, facial recognition technology has been widely criticized for higher error rates in identifying people of color, leading to wrongful arrests and discriminatory practices. In hiring algorithms, biases in data sets can result in unfair treatment of candidates based on race, gender, or socioeconomic status reinforcing inequality rather than addressing it.

2. **Privacy Concerns**: The deployment of AI technologies often involves the collection of massive amounts of personal data. Every click, purchase, and conversation become a data point, fueling AI systems that can predict behavior, preferences, and even health conditions. But what happens to our right to privacy in this AI-driven world? Who owns this data, and how is it being used? As companies and governments increasingly rely on AI for surveillance and control, individuals may find themselves constantly monitored, their lives reduced to algorithms. The erosion of privacy risks turning society into a digital panopticon where every move is watched.

3. **Autonomous Decision-Making**: As machines take on more complex decision-making tasks, we face a moral dilemma: can we trust AI to make ethical decisions on behalf of humans? For example, in the case of autonomous vehicles, who should a self-

driving car prioritize in an unavoidable accident? These life-and-death decisions, once the domain of human judgment, are now being programmed into machines, raising unsettling questions about who is responsible when something goes wrong.

The Potential for Misuse

AI and robotics, like any powerful technology, are tools that can be used for good or for harm. The potential for misuse is vast, and if left unchecked, these technologies could be weaponized in ways that threaten society.

1. **Weaponization of AI**: In the military sphere, AI has the potential to revolutionize warfare—but not without raising grave ethical concerns. Autonomous weapons systems, often dubbed "killer robots," could make decisions about life and death without any human oversight. These systems could be programmed to target enemies or even civilians based on predefined criteria, but what happens when the algorithm makes a mistake? The introduction of AI into warfare risks creating scenarios where machines, not humans, dictate the outcome of conflicts, leading to catastrophic consequences.

2. **Deepfakes and Misinformation**: The rise of deepfake technology, which uses AI to create hyper-realistic but entirely fabricated videos and audio, poses a growing threat to information integrity. As this technology becomes more sophisticated, the line between fact and fiction blurs, making it increasingly difficult for the public to trust what

they see and hear. Deepfakes can be weaponized to manipulate public opinion, spread disinformation, or tarnish reputations, threatening the very foundations of democracy and societal trust.

3. **Job Displacement and Economic Inequality**: Perhaps the most immediate concern surrounding AI and robotics is the displacement of human workers. Automation is rapidly transforming industries like manufacturing, transportation, and retail, replacing jobs that have historically been a foundation of middle-class livelihoods. As robots take over repetitive tasks, millions of people could find themselves unemployed and unemployable in an economy that increasingly values high-tech skills over manual labor. The result? A widening gulf between those who benefit from technological advancement and those left behind.

The Societal Consequences

Beyond individual risks, the widespread adoption of AI and robotics could lead to profound societal shifts, many of which may not be entirely positive.

1. **Erosion of Human Agency**: As AI takes over more decision-making responsibilities, there is a danger that individuals will become increasingly dependent on machines. We may lose the ability to think critically, solve problems independently, or even make moral choices, as we allow algorithms to guide us. The convenience of AI could come at the cost of our own agency, reducing us to passive participants in a world run by machines.

2. **Surveillance and Control**: The integration of AI into surveillance technologies poses serious risks to civil liberties. Governments and corporations can now monitor citizens on an unprecedented scale,

using AI to track movements, predict behavior, and even suppress dissent. In a world where AI-powered surveillance is ubiquitous, we may face a future where individuals are constantly monitored, and personal freedom becomes a relic of the past.

3. **Social Isolation**: As robots are increasingly used in roles traditionally filled by humans —caregiving, companionship, customer service— there is a risk that society could become more isolated. While robotic caregivers may offer convenience, they cannot replace the emotional and psychological connection that comes from human interaction. A world where people rely on robots for companionship may lead to increased social isolation, loneliness, and the degradation of community bonds.

Conclusion

The dark side of AI and robotics presents a stark reminder that technological advancement, while exciting, comes with significant risks. We must approach these innovations with caution and foresight, ensuring that ethical considerations, accountability, and transparency are prioritized. AI and robotics will shape the future, but it is up to us to determine whether that future is one where technology serves humanity or controls it.

In the chapters that follow, we will dive deeper into these risks, examining the specific threats posed by AI and robotics, and exploring how we can navigate the complex ethical landscape of this new technological frontier.

CHAPTER 3: THE RISE OF ARTIFICIAL INTELLIGENCE: FROM PROMISE TO PERIL

Artificial Intelligence (AI) has rapidly evolved from a theoretical concept to an integral part of our daily lives. As AI systems become increasingly sophisticated, they promise to enhance efficiency, solve complex problems, and improve decision-making across various domains. However, this rise also brings with it a host of challenges and risks that must be acknowledged and managed. In this chapter, we will explore the journey of AI, its transformative potential, and the perils that accompany its advancement.

The Evolution of AI Technology

The development of AI can be traced back to the mid-20th century, with initial theories and models focused on mimicking human cognition. Early breakthroughs in machine learning and neural networks laid the groundwork for today's advanced AI systems. The key milestones in AI evolution include:

1. **Machine Learning and Deep Learning**: Machine learning algorithms enable computers to learn from data without explicit programming. Deep learning, a subset of machine learning, employs neural networks to analyze vast datasets, leading to significant advancements in image recognition, natural language processing, and autonomous systems. These technologies have powered applications from virtual assistants to self-driving cars.

2. **Natural Language Processing (NLP)**: NLP allows machines to understand, interpret, and respond to human language. Applications such as chatbots and translation services have improved communication and accessibility. However, the complexity of language and context can lead to misunderstandings and miscommunications, raising concerns about the reliability of AI in critical settings.

3. **Autonomous Systems**: The rise of autonomous systems, including drones, robots, and self-driving vehicles, represents a significant leap in AI capabilities. These technologies have the potential to revolutionize industries such as transportation, logistics, and healthcare. Yet, the deployment of autonomous systems also raises questions about accountability, safety, and ethical considerations.

The Promises of AI

The potential benefits of AI are vast, offering solutions to some of humanity's most pressing challenges:

1. **Healthcare Innovations**: AI is revolutionizing

healthcare through predictive analytics, personalized medicine, and improved diagnostics. Machine learning algorithms can analyze medical data to identify patterns, enabling earlier detection of diseases and more effective treatment plans.

2. **Efficiency and Automation**: AI-driven automation can streamline processes across industries, leading to increased productivity and cost savings. For instance, AI algorithms can optimize supply chain management, reducing waste and improving resource allocation.

3. **Environmental Solutions**: AI can contribute to addressing climate change and environmental degradation. From optimizing energy consumption to monitoring deforestation and pollution, AI technologies hold promise in creating more sustainable practices.

The Perils of AI

Despite its potential, the rapid rise of AI comes with significant risks that cannot be overlooked:

1. **Job Displacement**: As AI systems automate tasks traditionally performed by humans, job displacement becomes a pressing concern. While new jobs may be created in tech sectors, many workers in industries such as manufacturing and transportation face the threat of redundancy. This shift could exacerbate economic inequality and lead to social unrest.

2. **Security Risks**: The integration of AI into critical infrastructure and cybersecurity poses significant risks. As AI systems become more sophisticated, they also become targets for malicious actors. AI-driven cyberattacks could lead to data breaches, financial losses, and disruptions to essential

services.

3. **Unintended Consequences**: The complexity of AI systems can lead to unintended consequences that are difficult to predict. For example, biased algorithms can perpetuate discrimination in hiring, lending, and law enforcement. These unintended effects highlight the need for transparency and accountability in AI development.

4. **Ethical Dilemmas**: The deployment of AI raises ethical questions about the limits of machine decision-making. For instance, in autonomous vehicles, how should the AI respond in situations where a crash is unavoidable? The moral implications of programming these decisions demand careful consideration.

Navigating the Future of AI

To harness the benefits of AI while mitigating its risks, a collaborative approach involving governments, industries, and society is essential:

1. **Establishing Regulatory Frameworks**: Governments must develop clear regulations to guide the ethical development and deployment of AI technologies. These frameworks should prioritize transparency, accountability, and public safety.

2. **Promoting Public Awareness**: Educating the public about AI, its benefits, and its risks is crucial for informed decision-making. An informed citizenry can better advocate for ethical practices and hold stakeholders accountable.

3. **Encouraging Ethical AI Development**: Organizations involved in AI research and development should prioritize ethical considerations at every stage of the process.

Collaboration between technologists, ethicists, and sociologists can foster responsible innovation.

Conclusion

The rise of artificial intelligence represents a transformative shift in how we live and work. While its potential benefits are immense, the associated risks must be taken seriously. By fostering a responsible approach to AI development, we can navigate the fine line between harnessing its power and preventing its pitfalls. In the subsequent chapters, we will delve deeper into the specific dangers posed by AI and robotics, including their implications for privacy, security, and social cohesion.

CHAPTER 4: THE SECRETS OF AI AND ROBOTS

The Hidden Complexity of AI

Artificial Intelligence, once the realm of science fiction, has rapidly integrated into our daily lives. From virtual assistants like Siri and Alexa to complex algorithms powering social media feeds and autonomous vehicles, AI has reshaped the fabric of modern society. However, the complexities and nuances of AI technology often remain shrouded in mystery, leaving many unaware of its potential consequences.

One of the most significant secrets of AI lies in its learning mechanisms. Most AI systems use deep learning, a subset of machine learning that mimics the way human brains operate. These systems analyze vast amounts of data, identifying patterns and making predictions. However, the vast datasets on which they are trained can inadvertently encode biases present in society. If these biases go unchecked, they can lead to discriminatory outcomes, impacting hiring practices, law enforcement, and even healthcare decisions.

For example, an AI system trained on historical hiring data may replicate past biases, inadvertently favoring certain demographics over others. This highlights the importance of scrutinizing the data used to train AI systems and implementing measures to mitigate bias, ensuring that AI operates equitably.

Robotics: The Unseen Risks

Robots, too, have become a cornerstone of technological advancement. From manufacturing lines to surgery rooms, robots improve efficiency and precision. Yet, as robots become more sophisticated, the risks associated with their deployment increase. For instance, a malfunctioning robot in a manufacturing plant can cause serious injury or even death. The challenge lies in the unpredictability of these machines. Unlike traditional machinery, robots equipped with AI can adapt and learn from their environments, which can lead to unintended behaviors.

Moreover, the potential for autonomous robots to operate without human oversight raises ethical dilemmas. Imagine a scenario where an autonomous delivery drone mistakenly identifies a person as a threat and takes evasive action. Such incidents could lead to serious legal and ethical questions about accountability and the decision-making processes of AI systems. This lack of oversight necessitates a robust framework for the ethical deployment of robotic technologies, ensuring that humans remain in the loop, particularly in critical applications.

The Dark Side of Automation

The rise of AI and robotics brings with it the very real fear of job displacement. As machines become capable of performing tasks traditionally handled by humans, entire industries are at risk of upheaval. This automation can lead to economic disparities, where the benefits of increased productivity and efficiency accrue to a small elite while large segments of the workforce find themselves without jobs.

The challenge is compounded by the fact that many of the jobs most susceptible to automation are those that require low skill levels, often held by vulnerable populations. The economic divide could widen further, leading to societal unrest as displaced workers struggle to find new opportunities in an increasingly automated world. It is crucial for policymakers to address these challenges by investing in retraining programs and educational initiatives that equip workers with the skills needed for the jobs of the future.

Secrets of AI Ethics

As AI continues to evolve, the question of ethical AI becomes increasingly critical. Who is responsible when an AI system makes a mistake? How do we ensure that these systems operate fairly and transparently? The algorithms that govern AI behavior often remain opaque, even to their creators. This lack of transparency can lead to situations where decisions are made without human oversight, raising concerns about accountability.

Moreover, as AI systems collect and analyze more personal data, issues of privacy and consent come to the forefront. Many users remain unaware of the extent to which their data is collected, analyzed, and utilized by AI systems. The potential for misuse of this data—whether by corporations or governments—poses a significant threat to individual freedoms and privacy. To address these issues, it is essential to establish clear regulations governing data collection, usage, and user consent, ensuring that individuals retain control over their personal information.

The Future: A Tipping Point

As we advance further into the era of AI and robotics, we stand at a critical juncture. The decisions we make today regarding the development, regulation, and deployment of these technologies will shape the future of society. Will we harness the potential of AI to benefit all, or will we allow it to exacerbate existing inequalities and create new ethical dilemmas?

In the coming chapters, we will delve deeper into the implications of AI on society, exploring potential regulations and frameworks needed to guide ethical AI development. We will also examine the role of humans in this new landscape and how we can prepare for a future where AI and robotics play a central role in our lives. By taking a proactive approach to AI and robotics, we can create a future that not only embraces innovation but also safeguards our values and humanity.

CHAPTER 5: THE HUMAN-MACHINE RELATIONSHIP

A New Era of Interaction

As technology advances at an unprecedented pace, the relationship between humans and machines is evolving into something profoundly complex. No longer are machines mere tools; they are becoming integral partners in our daily lives, capable of learning, adapting, and even mimicking human behaviors. This shift raises crucial questions about how we interact with these technologies and what it means for our identity, relationships, and society.

Consider the impact of AI-driven virtual assistants. These programs not only perform tasks but also engage in conversations that simulate human interaction. While they can provide convenience and efficiency, they can also create a false sense of companionship. Many people now confide in their virtual assistants, sharing personal thoughts and feelings. This emotional engagement with technology can lead to loneliness in real-life interactions, as individuals might prefer conversing with a machine rather than engaging with fellow humans. The comfort of a non-judgmental digital companion can unintentionally foster isolation, making it essential to examine how technology influences our social skills and emotional health.

The Emotional Connection to Technology

The rise of social robots' machines designed to interact with humans on a social level further complicates our relationship with technology. Robots like Sophia, an AI-powered humanoid, have gained media attention for their ability to engage in conversation and display human-like emotions. As these machines become more sophisticated, they raise profound ethical and psychological questions. Can we form genuine emotional bonds with robots? What happens when individuals develop feelings for machines that are not capable of reciprocal emotion?

The phenomenon of emotional attachment to robots has been observed in various settings, from elderly care facilities where social robots provide companionship to children who form friendships with interactive toys. While these relationships can offer comfort and alleviate loneliness, they also pose risks. Dependency on machines for emotional support can diminish the quality of human relationships and lead to an unhealthy reliance on technology for emotional well-being. For instance, a child who prefers a robotic friend over human companionship may miss critical social development opportunities, highlighting the need for balance in how we incorporate technology into our lives.

The Role of AI in Decision-Making

As machines become more integrated into our lives, they are increasingly taking on roles traditionally held by humans. From algorithm-driven hiring processes to AI systems that diagnose medical conditions, the reliance on machines for decision-making is becoming the norm. This shift raises critical questions about accountability and bias. When an AI system makes a decision that negatively impacts a person's life, who is responsible? The developer? The company? Or the machine itself?

Furthermore, the potential for bias in AI systems is a pressing concern. If algorithms are trained on flawed data, they can perpetuate existing prejudices, leading to unfair outcomes. For example, an AI hiring tool trained on data from a company with a history of gender bias may inadvertently favor male candidates over equally qualified female candidates. This situation underscores the necessity for ethical guidelines in AI development, ensuring that the decisions made by machines reflect fairness and equity. To mitigate these risks, organizations must implement robust auditing processes that continuously assess AI performance and address biases promptly.

The Fear of Dehumanization

As we cede more control to machines, a growing fear is that we may lose elements of our humanity. The efficiency and speed offered by technology often come at the cost of personal connections and empathy. In workplaces where AI automates tasks, there's a risk that human oversight becomes diminished, leading to a culture where productivity trumps personal interaction. The challenge lies in finding a balance between leveraging technology for efficiency and preserving the human touch that defines our interactions.

Moreover, the infiltration of AI into personal relationships can create a divide. For example, couples may find themselves competing for attention with smart devices. As people become more engrossed in their screens, the quality of face-to-face interactions may decline, leading to feelings of isolation even in the presence of others. It is vital for individuals and families to set boundaries around technology use, fostering environments where human connections remain a priority.

The Future of the Human-Machine Relationship

Looking ahead, the human-machine relationship will likely continue to evolve in ways we cannot fully predict. As we integrate AI and robotics into our daily lives, we must confront the implications of this dependence. Are we designing machines to serve us, or are we inadvertently creating a reality where technology dictates our interactions and decisions?

To navigate this future, we must establish ethical frameworks that prioritize human dignity and promote responsible use of technology. This includes fostering a culture that values human connections while embracing the benefits of innovation. As we explore the potential of AI and robotics, we must remain vigilant, ensuring that our relationship with technology enhances our humanity rather than diminishes it. Engaging in open discussions about the role of technology in our lives and advocating for responsible design will be essential as we move toward an increasingly automated future.

CHAPTER 6: THE ETHICAL IMPLICATIONS OF AI AND ROBOTICS

The Ethical Landscape of Artificial Intelligence

As artificial intelligence and robotics continue to advance, they bring not only groundbreaking innovations but also profound ethical dilemmas. These technologies have the potential to revolutionize industries and improve our lives, yet they also pose significant challenges that must be addressed. This chapter explores the ethical landscape surrounding AI and robotics, examining issues such as privacy, bias, accountability, and the moral status of machines.

Privacy Concerns in a Digital Age

One of the most pressing ethical issues related to AI is privacy. As AI systems increasingly collect and analyze vast amounts of personal data, the potential for misuse becomes a critical concern. For example, social media platforms and smart devices often track user behavior to deliver personalized content and advertisements. While this can enhance user experience, it also raises questions about consent and the extent to which individuals are aware of how their data is being used. The pervasive nature of data collection has made it challenging for individuals to understand the implications of their digital footprints, often leading to a false sense of security regarding their privacy.

Moreover, facial recognition technology, employed in various public and private sectors, poses significant privacy challenges. While it can enhance security, its deployment often occurs without public knowledge or consent. This lack of transparency can lead to surveillance overreach, creating a society where individuals are constantly monitored. The ethical dilemma arises: how do we balance security with the right to privacy? Establishing clear regulations and guidelines for data usage is essential to protect individual privacy in the age of AI. This includes implementing measures that require explicit user consent, allowing individuals to understand what data is collected and how it is used.

Bias and Discrimination in Algorithms

Bias in AI algorithms is another critical ethical concern. AI systems are trained on data sets that reflect historical patterns and societal norms. If these data sets are biased, the algorithms can perpetuate and even exacerbate existing inequalities. For instance, AI used in hiring processes may unintentionally favor certain demographics, leading to discrimination against underrepresented groups. In some cases, AI systems have been found to exhibit racial or gender bias, raising alarm bells about their application in sensitive areas like employment, lending, and criminal justice.

The ethical implications of biased AI extend to law enforcement as well. Predictive policing algorithms, which analyze data to forecast crime, can disproportionately target marginalized communities, perpetuating cycles of discrimination and inequality. This situation raises significant ethical questions: How do we ensure fairness in AI systems? What accountability measures can be implemented to prevent harm caused by biased algorithms? Addressing these issues requires a commitment to transparency, fairness, and ongoing evaluation of AI systems. Incorporating diverse datasets, engaging with affected communities, and conducting regular audits can help mitigate biases and enhance the fairness of AI applications.

Accountability and Responsibility

As AI systems take on increasingly autonomous roles, questions of accountability become paramount. When an AI makes a decision that leads to negative consequences—such as a self-driving car accident—who is responsible? The manufacturer, the programmer, or the user? This ambiguity complicates the notion of accountability and raises ethical concerns about the development and deployment of autonomous systems. In the absence of clear accountability, victims of AI-related harm may find it challenging to seek justice or compensation.

To address this challenge, ethical frameworks must be established that clearly define the responsibilities of developers, companies, and users. Engaging in discussions about accountability will help shape the development of regulations that protect individuals and communities from harm caused by AI systems. Some propose the creation of "AI liability insurance" models to ensure that developers and companies are held financially accountable for the actions of their systems. By promoting accountability, we can foster a culture of responsibility in AI development.

The Moral Status of Machines

The question of the moral status of AI and robots is a philosophical and ethical conundrum that is gaining prominence. As machines become more sophisticated, they increasingly exhibit traits associated with consciousness, such as learning, adaptation, and emotional responses. This development raises profound questions: Should machines have rights? Can they be held morally accountable for their actions?

While some argue that machines lack consciousness and cannot possess moral status, others contend that the capabilities of AI systems necessitate a reevaluation of our ethical frameworks. As we develop more advanced AI, we must consider how we define personhood and what moral considerations should be extended to machines. This exploration will be crucial in shaping the future of human-machine interactions. Moreover, as we navigate these questions, we must remain cautious of anthropomorphizing machines and conflating their capabilities with human-like qualities, which may lead to misplaced trust and ethical oversights.

Balancing Innovation with Ethical Responsibility

The ethical implications of AI and robotics present significant challenges, but they also offer opportunities for thoughtful reflection and improvement. As we navigate the complexities of this technological landscape, it is essential to prioritize ethical considerations alongside innovation.

1. **Develop Ethical Guidelines**: Industry stakeholders, researchers, and policymakers should collaborate to establish ethical guidelines that govern the development and deployment of AI technologies. These guidelines should prioritize transparency, accountability, and fairness in AI systems. Additionally, ethical considerations should be integrated into the design process from the outset, rather than being an afterthought.

2. **Foster Ethical Literacy**: Educating developers, engineers, and the public about the ethical implications of AI is vital. By fostering ethical literacy, we can encourage responsible decision-making in the design and implementation of AI

technologies. Initiatives aimed at raising awareness about the potential risks and ethical dilemmas of AI can empower individuals to advocate for responsible technology use.

3. **Engage Diverse Perspectives**: The development of AI should involve diverse voices and perspectives, ensuring that a wide range of ethical considerations is taken into account. Engaging ethicists, social scientists, and community representatives will help create more equitable and inclusive AI systems. Collaborative efforts can bridge gaps in understanding and foster innovative solutions that prioritize human well-being.

4. **Promote Ethical AI Research**: Funding and supporting research initiatives that focus on the ethical implications of AI and robotics is crucial. This research can provide insights into best practices and inform the development of responsible technologies. Academic institutions and industry leaders must work together to create interdisciplinary research programs that examine the social, ethical, and technical dimensions of AI.

As we advance into an era dominated by AI and robotics, it is imperative to confront the ethical challenges they present. By prioritizing ethical considerations in the design and implementation of these technologies, we can ensure that innovation serves the greater good, fostering a future where humanity thrives alongside intelligent machines.

CHAPTER 7: THE SOCIAL RAMIFICATIONS OF AI AND ROBOTICS

The Changing Nature of Work

As AI and robotics become increasingly integrated into various industries, the nature of work is evolving rapidly. While these technologies have the potential to enhance productivity and efficiency, they also pose significant challenges, particularly regarding job displacement. Automation is transforming traditional jobs, replacing tasks that were once performed by humans, and leading to concerns about widespread unemployment.

The impact of automation is already visible in sectors such as manufacturing, transportation, and retail. For instance, self-checkout kiosks in supermarkets and automated assembly lines in factories have replaced many cashier and assembly jobs. Additionally, advancements in self-driving technology threaten to displace millions of truck drivers, delivery personnel, and taxi operators. The World Economic Forum has predicted that by 2025, 85 million jobs may be displaced by the shift to automation, while 97 million new roles could emerge. However, the transition is not always smooth, and many workers may find it challenging to adapt to the changing job landscape.

To effectively navigate this transition, a nuanced understanding of the evolving job market is essential. It is not only the quantity of jobs that is changing but also their nature. Many of the new roles emerging will likely demand higher levels of education and technical expertise, leaving low-skilled workers vulnerable. This shift necessitates a collective effort to develop adaptable workforce strategies that can respond to the demands of a technology-driven economy.

The Skills Gap and Reskilling Efforts

The shift toward automation raises critical questions about the skills workers will need to thrive in a technology-driven economy. As routine tasks are increasingly performed by machines, there is a growing demand for workers with advanced technical skills, critical thinking, and creativity. However, many workers lack access to the training and resources necessary to develop these skills, leading to a significant skills gap.

Reskilling initiatives are essential to address this challenge. Companies, governments, and educational institutions must collaborate to provide accessible training programs that equip workers with the skills needed for the jobs of the future. Programs that focus on STEM (science, technology, engineering, and mathematics), digital literacy, and soft skills will be crucial in preparing the workforce for an increasingly automated landscape.

Moreover, businesses can play a pivotal role by investing in their employees' ongoing education. Offering in-house training programs or partnerships with local educational institutions can help workers transition smoothly into new roles. The focus should also extend beyond technical skills; fostering critical soft skills such as emotional intelligence, adaptability, and problem-solving will be vital in ensuring a resilient workforce.

Social Inequality and Job Displacement

The rapid adoption of AI and robotics is exacerbating existing social inequalities. While technology can create new job opportunities, it often benefits those who are already skilled and educated, leaving marginalized groups at a disadvantage. As high-skilled workers thrive in the tech-driven economy, low-skilled workers may struggle to find stable employment, leading to increased economic disparities.

Moreover, the concentration of wealth and power in the hands of a few tech giants raises ethical concerns about the societal implications of automation. The companies that dominate the AI landscape often prioritize profit over social responsibility, leading to a situation where the benefits of technology are not equitably distributed. This growing divide can foster resentment and social unrest, as those left behind by technological advancements feel neglected and powerless.

To mitigate these effects, targeted interventions are necessary. Policymakers should consider implementing progressive taxation and wealth redistribution measures to address the inequalities exacerbated by automation. Moreover, initiatives aimed at supporting entrepreneurship in underrepresented communities can empower individuals to create their own opportunities within a shifting job landscape.

Impact on Interpersonal Relationships

The rise of AI and robotics is not only transforming the workplace but also influencing how individuals interact with one another. As machines take on more tasks, from customer service roles to companionship, the dynamics of human relationships are shifting. While AI-powered chatbots and virtual assistants offer convenience, they may also erode the quality of interpersonal interactions.

The proliferation of technology in daily life can lead to social isolation, as individuals become more reliant on machines for communication and companionship. This shift raises important questions about the role of human connection in a world increasingly dominated by AI. Are we sacrificing meaningful relationships for efficiency? How do we ensure that technology enhances, rather than replaces, human interaction?

To counteract potential social isolation, it is crucial to design technology with a human-centric approach. Innovations that promote social connection and collaboration can foster community rather than diminish it. For instance, platforms that facilitate genuine interaction—such as virtual reality environments designed for social engagement—can help maintain human relationships in an increasingly automated world.

The Ethical Dilemmas of AI in Society

The societal ramifications of AI and robotics also bring forth ethical dilemmas that must be addressed. As these technologies become more embedded in our lives, decisions about their use often reflect societal values and priorities. For instance, the use of AI in surveillance systems raises questions about privacy and civil liberties. How do we balance security needs with the right to privacy?

Furthermore, the potential for AI to reinforce existing biases highlights the ethical responsibility of developers and policymakers. If AI systems are trained on biased data, they may perpetuate discrimination and inequality. Ensuring that AI is developed and deployed in a manner that is fair and equitable requires a commitment to ethical practices and oversight.

The implementation of ethical review boards that include diverse stakeholders can help ensure that AI applications are scrutinized from multiple perspectives. Engaging communities in discussions about AI development can foster trust and ensure that technological advancements reflect the needs and values of society at large.

Conclusion: Navigating the Social Landscape of AI and Robotics

The social ramifications of AI and robotics are complex and multifaceted. As we navigate this evolving landscape, it is essential to consider the implications of these technologies on work, social structures, and interpersonal relationships. By prioritizing ethical considerations, reskilling efforts, and inclusive practices, we can work toward a future where technology enhances human life rather than diminishes it.

1. **Emphasize Inclusive Policies**: Governments and organizations should develop policies that promote inclusivity and equity in the face of automation. This includes support for reskilling programs and access to education for marginalized communities.

2. **Encourage Human-Centric AI**: As we design and implement AI systems, we must prioritize human values and social impact. Technology should enhance human connection and collaboration rather than replace it.

3. **Foster Open Dialogue**: Engaging in conversations about the societal implications of AI and robotics is crucial. Stakeholders from various sectors should come together to discuss the challenges and opportunities posed by these technologies.

4. **Monitor and Evaluate Impact**: Ongoing evaluation of the social impact of AI and robotics is essential. Researchers, policymakers, and industry leaders should work together to assess the effects of these technologies and make necessary adjustments.

As we continue to integrate AI and robotics into our lives, we must remain vigilant about their social implications. By approaching these technologies with a critical lens, we can navigate the challenges they present while maximizing their potential to improve the human experience.

CHAPTER 8: THE POTENTIAL DANGERS OF AI AND ROBOTICS

The Unintended Consequences of Automation

As AI and robotics permeate various aspects of our lives, the potential dangers become increasingly apparent. While these technologies promise efficiency and convenience, they also come with a host of unintended consequences that could have far-reaching effects on society. One of the most pressing concerns is job displacement and the resulting economic instability. As robots take over tasks traditionally performed by humans, entire industries may face upheaval, leading to widespread unemployment and a skills gap that leaves many workers behind.

Furthermore, the shift towards automation often favors certain sectors over others, leading to regional disparities. Areas heavily reliant on industries susceptible to automation may experience significant economic downturns, exacerbating poverty and social instability. Addressing these issues requires proactive policies that anticipate these shifts and support affected communities.

Additionally, the introduction of AI into decision-making processes raises significant ethical concerns. Algorithms, which are often opaque and difficult to understand, may perpetuate biases inherent in the data used to train them. For example, AI systems employed in hiring processes or criminal justice may inadvertently discriminate against certain demographic groups, leading to a cycle of inequality and injustice. The reliance on automated systems necessitates rigorous oversight and accountability mechanisms to ensure fairness and transparency.

The Risk of Autonomous Weapons

One of the most alarming aspects of advancing robotics is the development of autonomous weapons systems. These AI-driven technologies can operate without human intervention, making decisions about life and death based on algorithms. The implications of such technologies are profound and disturbing.

The potential for autonomous weapons to engage in warfare without human oversight raises questions about accountability and ethics. If an AI system miscalculates a threat or acts on flawed data, who is responsible for the consequences? The erosion of human agency in life-and-death decisions could lead to catastrophic outcomes, making it imperative to establish strict regulations governing the use of autonomous weapons.

Furthermore, the proliferation of autonomous weaponry could lower the threshold for engaging in conflict, making wars more likely and potentially more devastating. The international community must engage in dialogue and negotiations to establish treaties that limit the development and use of these technologies.

Security Threats and Cyber Vulnerabilities

The integration of AI and robotics into critical infrastructure and national security systems also exposes vulnerabilities that could be exploited by malicious actors. Cybersecurity threats are a growing concern as our reliance on technology increases. Hackers could target AI systems, manipulating them to create chaos or disruption in essential services.

For instance, AI algorithms used in power grids, transportation systems, or financial institutions are susceptible to hacking. A successful cyberattack on these systems could result in catastrophic consequences, from widespread blackouts to financial collapse. The potential for AI to be weaponized in cyber warfare highlights the urgent need for robust security measures and ethical considerations in the development of AI technologies.

Moreover, as AI becomes integral to national defense strategies, the arms race in AI technology could escalate geopolitical tensions. Countries may prioritize rapid AI advancements over ethical considerations, leading to a precarious global landscape. International cooperation is essential to establish norms and regulations that mitigate these risks.

The Erosion of Privacy and Surveillance

AI's capabilities in data analysis and facial recognition technology have led to an increase in surveillance practices. Governments and corporations are using AI to monitor individuals, raising significant concerns about privacy and civil liberties. The pervasive use of surveillance technology can create a society where individuals are constantly watched, leading to self-censorship and a chilling effect on free speech.

Moreover, the collection of vast amounts of personal data by AI systems poses ethical dilemmas regarding consent and data ownership. Individuals may not fully understand how their data is being used or the implications of its collection, leading to a loss of control over personal information. It is essential to develop regulations that ensure transparency in data collection practices and empower individuals to control their information.

The potential misuse of surveillance technologies by authoritarian regimes further underscores the need for vigilant oversight. Advocating for privacy rights and ethical standards in technology deployment is critical in safeguarding democratic values.

The Psychological Impact of AI

The rise of AI and robotics also presents psychological challenges that warrant consideration. As humans interact more frequently with machines, the potential for emotional detachment and social isolation increases. For example, AI companions designed to provide emotional support may lead individuals to prefer interactions with machines over genuine human connections.

The impact of AI on mental health is a growing area of concern. Studies suggest that excessive reliance on technology for companionship can lead to feelings of loneliness and depression. As we embrace AI-driven solutions to enhance our lives, it is essential to remain vigilant about the potential for these technologies to negatively affect our emotional well-being.

Additionally, the pervasive presence of AI in daily life may exacerbate anxiety related to job security and societal relevance. As automation replaces certain tasks, individuals may grapple with existential questions about their purpose and identity in a rapidly changing world. Mental health resources must evolve to address these challenges, providing support for individuals navigating the complexities of an AI-driven society.

Conclusion: Navigating the Risks of AI and Robotics

The potential dangers of AI and robotics are multifaceted, touching upon ethical, security, privacy, and psychological dimensions. As society continues to advance technologically, it is crucial to address these risks head-on.

1. **Implement Ethical Guidelines**: Establishing clear ethical frameworks for the development and deployment of AI technologies is essential to ensure that they are used responsibly and do not perpetuate existing biases or inequalities.

2. **Enhance Cybersecurity Measures**: As AI systems become integral to critical infrastructure, investing in robust cybersecurity measures is paramount. Governments and organizations must collaborate to protect these systems from potential threats.

3. **Foster Open Discussions**: Engaging in open conversations about the societal implications of AI and robotics will help create a more informed public. Policymakers, technologists, and citizens must work together to address the challenges posed by these technologies.

4. **Promote Human-Centric Design**: Developing AI systems with a focus on enhancing human agency and well-being will be essential in mitigating the potential negative impacts. Designers and developers should prioritize user needs and ethical considerations in their work.

By addressing the potential dangers of AI and robotics thoughtfully and proactively, we can navigate the complexities of these technologies while maximizing their benefits for society. A collaborative approach involving diverse stakeholders is crucial in shaping a future where AI and robotics contribute positively to human life.

CHAPTER 9: THE ETHICAL CHALLENGE OF DISRUPTIVE TECHNOLOGIES

Introduction: The Rapid Pace of Innovation
As we continue to witness the rapid advancement of technology, particularly in AI and robotics, ethical challenges become increasingly prominent. These disruptive technologies have the power to reshape industries, redefine societal norms, and alter human experiences. However, with great power comes great responsibility. This chapter will delve into the ethical considerations surrounding these technologies, focusing on issues such as accountability, bias, transparency, and the potential impact on human rights.

The Accountability Conundrum
One of the most pressing ethical challenges posed by AI and

robotics is accountability. When autonomous systems make decisions, who is held responsible for the outcomes? For instance, if an autonomous vehicle is involved in an accident, is it the manufacturer, the software developer, or the owner who is liable?

This ambiguity complicates the legal landscape and raises questions about the moral responsibilities of those who create and deploy these technologies. Clear frameworks must be established to determine accountability in cases of malfunction or harm caused by AI systems. Failing to address this issue may result in a lack of trust in autonomous technologies and hinder their widespread adoption.

Bias in AI Systems

Another significant ethical challenge is the potential for bias in AI algorithms. AI systems are trained on historical data, which may reflect societal biases and inequalities. If these biases go unrecognized, the technologies could perpetuate and amplify existing prejudices in areas such as hiring practices, law enforcement, and credit scoring.

For example, facial recognition technology has been shown to exhibit racial and gender biases, leading to higher rates of misidentification for people of color and women. It is crucial for developers and organizations to prioritize fairness and inclusivity in AI training data and algorithms to mitigate these biases. Ethical AI design should aim to promote social equity rather than reinforce discrimination.

The Importance of Transparency

Transparency is a cornerstone of ethical technology development. As AI systems become more complex, understanding how they make decisions is essential for ensuring accountability and trust. However, many AI algorithms, particularly those based on deep learning, operate as "black boxes," making it difficult for users and stakeholders to comprehend their decision-making processes.

Promoting transparency involves providing clear explanations of how AI systems work and the data they rely on. By making AI algorithms more interpretable, organizations can foster public trust and encourage informed decision-making by users. This transparency is particularly vital in high-stakes situations, such as healthcare or criminal justice, where AI decisions can have profound consequences for individuals' lives.

Protecting Human Rights

The rise of AI and robotics also raises concerns about human rights. The integration of these technologies into surveillance systems poses a significant threat to privacy and civil liberties. Governments and corporations may misuse AI technologies for mass surveillance, leading to a loss of personal freedoms and the erosion of trust in public institutions.

Furthermore, the potential for AI to replace human jobs raises ethical concerns about economic rights. As automation increases, it is essential to consider how society will support those displaced by technology. Policymakers must prioritize creating safety nets and retraining programs to protect workers' rights and ensure equitable access to opportunities in an increasingly automated economy.

The Role of Ethics in Technology Design

To address these ethical challenges, a proactive approach to technology design is essential. Integrating ethical considerations into the development process ensures that AI and robotics align with societal values and human rights.

1. **Interdisciplinary Collaboration:** Bringing together experts from diverse fields, including ethicists, social scientists, and technologists, can foster a more comprehensive understanding of the ethical implications of AI. Collaborative efforts will lead to well-rounded solutions that consider various

perspectives.

2. **Establishing Ethical Guidelines:** Organizations and governing bodies should develop ethical guidelines and frameworks that outline the principles of responsible AI development. These guidelines should prioritize transparency, accountability, fairness, and respect for human rights.

3. **Public Engagement:** Engaging the public in discussions about the ethical implications of AI and robotics will promote informed consent and collective responsibility. By involving communities in the conversation, developers can better understand societal values and expectations.

Conclusion: Navigating the Ethical Landscape

The ethical challenges presented by AI and robotics require thoughtful consideration and action. As we navigate this complex landscape, it is crucial to prioritize accountability, address bias, ensure transparency, and protect human rights. By fostering a culture of ethical technology development, we can harness the potential of these disruptive technologies while mitigating their risks.

In this rapidly evolving field, the choices we make today will shape the future of AI and robotics. As leaders, developers, and stakeholders, we must commit to creating technologies that enhance the human experience, promote social equity, and uphold our shared ethical principles.

CHAPTER 10: THE INTERSECTION OF AI, ETHICS, AND HUMAN RIGHTS

Introduction: The Rise of AI and Human Rights
As artificial intelligence continues to permeate various aspects of daily life—from healthcare to law enforcement—the intersection of AI, ethics, and human rights becomes increasingly significant. The deployment of AI technologies raises critical questions about the protection of individual rights, privacy, and the potential for abuse. This chapter will explore the ethical dimensions of AI, examining how these technologies can both promote and threaten human rights.

AI in Surveillance and Privacy Rights
One of the most pressing human rights concerns associated

with AI is its use in surveillance systems. Governments and corporations are increasingly deploying AI-driven surveillance technologies, such as facial recognition and behavior analysis, in public spaces. While proponents argue that these technologies can enhance security and crime prevention, they also pose a significant threat to privacy rights and civil liberties.

1. **Invasive Surveillance:** The pervasive use of AI surveillance can lead to a society where individuals are constantly monitored, eroding the concept of privacy. This intrusion can stifle free expression and inhibit dissent, creating an environment of fear where people are reluctant to voice their opinions or engage in protests.

2. **Misuse of Data:** AI systems often rely on vast amounts of personal data to function effectively. Without stringent regulations and transparency, there is a risk that this data could be misused for purposes beyond its intended use, such as profiling individuals based on their behaviors or beliefs.

3. **Discriminatory Practices:** AI surveillance systems may disproportionately target marginalized communities, leading to a cycle of discrimination and social injustice. For instance, biased facial recognition algorithms have been shown to misidentify individuals from specific racial or ethnic backgrounds at higher rates, leading to unjust policing practices and systemic inequality.

The Right to Work in an Automated Future

The rapid advancement of AI technologies also raises concerns about the right to work and economic justice. As automation continues to replace traditional jobs, it is essential to consider the implications for workers' rights and economic stability.

1. **Job Displacement:** The integration of AI in various

sectors poses a risk of mass job displacement, particularly for low-skilled workers. This shift can exacerbate existing inequalities, leaving many individuals without stable employment or the means to support their families.

2. **Access to Retraining Programs:** To protect the right to work, it is vital to implement robust retraining programs and educational initiatives that empower displaced workers to transition into new roles. Ensuring equitable access to these resources is crucial for promoting economic justice and social mobility.

3. **Universal Basic Income:** Some experts argue that in an increasingly automated world, adopting policies such as Universal Basic Income (UBI) could provide a safety net for individuals affected by job displacement. By providing financial security, UBI can help safeguard individuals' rights to basic economic stability and dignity.

AI and Healthcare: Ethical Considerations

AI technologies hold great promise in the healthcare sector, from diagnostics to personalized medicine. However, the integration of AI in healthcare raises significant ethical considerations related to human rights and patient care.

1. **Bias in Medical Algorithms:** AI algorithms used in healthcare can perpetuate biases if they are trained on unrepresentative data. This can lead to disparities in treatment recommendations and outcomes, particularly for marginalized groups. Ensuring diversity in training data is essential for promoting equitable healthcare practices.

2. **Informed Consent and Autonomy:** As AI systems are increasingly used to make healthcare decisions,

the importance of informed consent becomes paramount. Patients must have the right to understand how AI technologies impact their care and to make choices based on that knowledge.

3. **Data Privacy in Healthcare:** The use of AI in healthcare necessitates the collection and storage of sensitive personal data. Protecting patient privacy and ensuring the ethical use of health data is critical to maintaining trust in healthcare systems.

The Global Implications of AI on Human Rights

The ethical challenges posed by AI are not confined to any single region; they have global implications. As AI technologies are developed and deployed across borders, it is essential to consider their impact on human rights worldwide.

1. **Digital Colonialism:** The global reach of AI technologies can lead to a form of digital colonialism, where wealthy nations impose their technologies and values on developing countries. This dynamic raises questions about sovereignty, cultural preservation, and the equitable distribution of technological benefits.

2. **International Regulations:** The lack of comprehensive international regulations governing AI raises concerns about the potential for abuse. Collaborative efforts among nations are necessary to establish ethical guidelines that protect human rights in the context of AI development and deployment.

3. **The Role of Civil Society:** Civil society organizations play a crucial role in advocating for human rights in the age of AI. By raising awareness, conducting research, and holding governments and corporations accountable, these organizations

can help ensure that technological advancements respect and promote human dignity.

Conclusion: A Call for Ethical Leadership

As AI continues to evolve and shape our world, the ethical considerations surrounding its development and use must remain at the forefront of discussions. Promoting a rights-based approach to AI ensures that these technologies enhance, rather than undermine, human rights and dignity.

Ethical leadership is crucial for navigating the complexities of AI and ensuring that advancements are aligned with societal values. By prioritizing accountability, transparency, and inclusivity, we can harness the potential of AI to create a more equitable and just future for all.

CHAPTER 11: THE PSYCHOLOGICAL IMPACT OF AI ON HUMAN BEHAVIOR

Introduction: AI as a Companion and a Tool
As artificial intelligence becomes an integral part of everyday life, it profoundly impacts how we think, interact, and perceive the world. From virtual assistants to social media algorithms, AI systems shape our experiences and influence our decisions. While these technologies can enhance convenience and efficiency, they also raise concerns about their psychological effects on individuals and society as a whole.

AI and Human Interaction

1. **Changing Communication Patterns:** AI-driven communication tools, such as chatbots and

virtual assistants, have transformed how we interact. While these technologies can facilitate communication, they may also lead to a reduction in face-to-face interactions, impacting social skills and emotional intelligence.

2. **The Rise of Social Isolation:** As people increasingly turn to AI for companionship or assistance, there is a risk of social isolation. Reliance on virtual interactions may diminish the quality and depth of human relationships, leading to feelings of loneliness and disconnection.

3. **Influence on Social Skills:** Children and adolescents growing up in an AI-rich environment may struggle with developing essential social skills. If interactions with peers are replaced by AI interactions, it could hinder their ability to navigate social cues and form meaningful connections.

The Emotional Impact of AI

1. **AI and Mental Health:** AI technologies can both help and hinder mental health. While AI-based applications provide support through therapy bots and mood tracking, they may also contribute to anxiety and depression if users rely too heavily on them instead of seeking human connection.

2. **The Dangers of Misinformation:** AI algorithms that curate news and social media feeds can create echo chambers, where users are exposed only to information that reinforces their existing beliefs. This can lead to increased polarization, anxiety, and fear, as individuals become more entrenched in their viewpoints.

3. **Fear of Obsolescence:** As AI systems become more capable, many individuals may experience anxiety about their job security and relevance in the

workforce. This fear of obsolescence can lead to stress and decreased self-esteem, impacting overall well-being.

The Cognitive Effects of AI

1. **Dependence on AI:** With the convenience of AI systems providing quick answers and assistance, there is a risk of cognitive laziness. Over-reliance on technology may diminish critical thinking skills and reduce our ability to solve problems independently.

2. **Information Overload:** AI technologies often curate vast amounts of information, which can lead to information overload. Individuals may struggle to filter relevant information, resulting in cognitive fatigue and difficulty making decisions.

3. **Behavioral Manipulation:** AI systems, particularly those employed in advertising and social media, can manipulate human behavior through targeted content and suggestions. This raises ethical concerns about autonomy and the extent to which AI can influence choices and preferences.

The Role of AI in Education

1. **Personalized Learning:** AI can revolutionize education by providing personalized learning experiences. Adaptive learning technologies can cater to individual students' needs, improving engagement and understanding. However, reliance on AI-driven education may also hinder the development of critical thinking and creativity.

2. **The Teacher-Student Dynamic:** As AI tools become more prevalent in classrooms, the role of teachers may shift. While AI can support educators, it is

crucial to maintain the human element in teaching, as relationships and emotional support are vital to effective learning.

Conclusion: Navigating the Psychological Landscape of AI

As we embrace the benefits of AI, it is essential to remain vigilant about its psychological impacts on individuals and society. By fostering awareness and encouraging responsible use of AI technologies, we can navigate the complexities of this new landscape while preserving the essential human qualities that define our interactions and relationships.

Moving forward, promoting mental well-being, enhancing social skills, and prioritizing emotional intelligence must remain at the forefront of discussions about AI's role in society. In an age increasingly shaped by artificial intelligence, ensuring that technology serves to enhance our humanity rather than diminish it is crucial.

CHAPTER 12: THE ROLE OF AI IN SHAPING PUBLIC OPINION AND BEHAVIOR

Introduction: The Power of AI in Communication

Artificial intelligence plays a pivotal role in shaping public opinion and behavior by influencing how information is disseminated and consumed. Through algorithms that curate content on social media platforms, news outlets, and search engines, AI has become a powerful force in determining what people see, read, and ultimately believe. This chapter delves into the mechanisms through which AI influences public opinion and the ethical considerations that arise from its use.

The Mechanisms of Influence

1. **Content Curation Algorithms:** AI algorithms analyze user behavior and preferences to curate personalized content. This tailored experience enhances engagement but can lead to filter bubbles, where individuals are exposed only to information that aligns with their existing beliefs. This narrowing of perspective can hinder critical thinking and reduce exposure to diverse viewpoints.

2. **Social Media Dynamics:** Platforms like Twitter, Facebook, and Instagram rely on AI-driven algorithms to determine what content goes viral. Posts that resonate with users—whether positive or negative—can gain traction quickly, influencing public sentiment on various issues. The rapid spread of misinformation and sensational content can significantly shape societal opinions.

3. **Echo Chambers:** AI's ability to create echo chambers exacerbates polarization. Users are more likely to engage with content that confirms their biases, leading to the reinforcement of existing beliefs. This can create divisions in society, as individuals become entrenched in their viewpoints and less willing to engage with opposing perspectives.

The Impact of Misinformation

1. **The Spread of Fake News:** AI can inadvertently facilitate the spread of misinformation by amplifying false narratives. Fake news stories often spread faster than factual reporting, creating confusion and distrust among the public. This phenomenon can undermine democracy and public

trust in institutions.

2. **Manipulation of Public Sentiment:** Malicious actors can use AI to manipulate public sentiment, employing tactics such as bot-driven campaigns to amplify divisive content. These tactics can create artificial trends, sway elections, and influence public policy by distorting the conversation around critical issues.

The Ethical Implications

1. **Responsibility of Tech Companies:** The responsibility for curbing the negative impacts of AI lies with technology companies that design these algorithms. As the architects of public discourse, they must navigate the ethical implications of their systems, balancing user engagement with the potential for harm.

2. **Transparency and Accountability:** Users should be aware of how AI algorithms shape their online experiences. Transparency in algorithmic processes can foster trust and enable individuals to make informed decisions about the information they consume. However, many companies are reluctant to disclose their methods, raising concerns about accountability.

3. **The Role of Regulation:** Governments and regulatory bodies must consider the implications of AI in shaping public opinion. Implementing regulations that promote ethical AI use, combat misinformation, and protect users' rights is essential in an increasingly digitized society.

The Future of Public Discourse

1. **AI as a Tool for Good:** While AI presents

challenges, it can also be harnessed for positive change. For instance, AI can help identify and flag misinformation, promote accurate reporting, and facilitate constructive dialogue across diverse perspectives. Embracing AI's potential for good requires a concerted effort from tech companies, policymakers, and society at large.

2. **Empowering Users:** Educating users about the influence of AI on public opinion is crucial. By fostering media literacy and critical thinking skills, individuals can better navigate the complexities of the information landscape, making informed choices about the content they engage with.

Conclusion: Navigating the AI-Driven Landscape

As AI continues to shape public opinion and behavior, it is vital to recognize its potential and the ethical challenges it poses. By promoting transparency, accountability, and responsible AI use, society can harness the power of artificial intelligence to foster informed public discourse and enhance democratic engagement.

In an era where information is readily available, understanding the role of AI in shaping opinions is essential. As individuals, we must remain vigilant and critically engage with the information we encounter, ensuring that our beliefs are informed by diverse perspectives rather than limited by algorithmic constraints.

CHAPTER 13: THE SOCIETAL IMPLICATIONS OF AI-DRIVEN AUTOMATION

Introduction: The Rise of Automation
Artificial intelligence and automation have revolutionized modern industries, changing how we work and live. While these technologies enhance efficiency and create new opportunities, they also pose significant societal implications that require careful consideration. This chapter explores how AI-driven automation reshapes the workforce, economy, and social dynamics.

The Transformation of the Workforce
1. **Job Displacement:** One major concern about automation is job displacement. As machines

increasingly perform tasks traditionally done by humans, workers in sectors like manufacturing and retail may lose their jobs. The World Economic Forum estimates millions could be affected, leading to economic uncertainty for many.

2. **New Job Creation:** Automation may eliminate some jobs but also creates new ones, particularly in AI, data science, and cybersecurity. The challenge is ensuring displaced workers can transition into these roles through reskilling and upskilling initiatives.

3. **The Need for Lifelong Learning:** The shift toward automation underscores the importance of lifelong learning. Workers must continuously adapt and acquire new skills to remain relevant. Collaboration among educational institutions, businesses, and governments is essential to provide accessible training programs.

Economic Implications

1. **Economic Inequality:** Automation could worsen existing economic inequalities. While skilled workers may thrive, those in low-skilled jobs might struggle to find employment, risking a widening wealth gap that threatens social stability.

2. **Productivity and Economic Growth:** On a positive note, AI-driven automation can enhance productivity and drive economic growth. Streamlining operations can lead to increased profitability, which, if managed responsibly, could create new wealth and opportunities for society.

3. **Impact on Small Businesses:** Small businesses may face unique challenges in adapting to automation. Larger corporations with the resources to invest in advanced technologies may gain a competitive edge,

leaving smaller enterprises behind. Supporting small businesses through grants and resources is vital for a diverse economy.

Social Dynamics and Relationships

1. **Changes in Work Culture:** Automation is reshaping work culture, making remote work more common. While this flexibility can improve work-life balance, it may also blur the lines between work and personal life, leading to potential burnout.

2. **Human Interaction and Connection:** The rise of automation risks diminishing human interaction. In sectors like healthcare and customer service, while AI can provide efficiency, it may compromise the essential empathetic touch that drives positive outcomes.

3. **Ethical Considerations:** The societal implications of AI-driven automation raise ethical questions regarding worker treatment and corporate responsibilities. Businesses must consider the impact of automated systems on employees and communities, ensuring equitable benefits from automation.

Conclusion: Navigating the Future of Work

The implications of AI-driven automation are complex. While increased efficiency and productivity offer promise, it is crucial to address the accompanying challenges and ethical concerns. Prioritizing reskilling, supporting small businesses, and fostering inclusive growth will allow society to harness automation's benefits while minimizing negative impacts.

As we navigate an increasingly automated future, adaptability and collaboration will be vital. We must embrace the opportunities AI presents while advocating for policies that prioritize social responsibility and well-being for all.

CHAPTER 14: THE ETHICAL DILEMMAS OF AI IN SOCIETY

Introduction: The Ethical Landscape of AI

As artificial intelligence becomes increasingly integrated into daily life, ethical dilemmas surrounding its use have emerged as critical societal concerns. While AI holds the potential to enhance efficiency, accuracy, and convenience, it raises profound questions about morality, fairness, and the implications of machine decision-making. This chapter explores the ethical challenges posed by AI technologies and the responsibilities that accompany their implementation.

The Problem of Bias in AI

1. **Inherent Biases:** A pressing ethical concern regarding AI is the presence of bias in algorithms. AI systems learn from historical data, and if that data

reflects societal biases, the resulting algorithms can perpetuate and even amplify these biases. For instance, facial recognition technologies often perform less accurately for individuals with darker skin tones, raising concerns about discrimination in law enforcement and hiring practices.

2. **Accountability for Bias:** The issue of accountability arises when AI systems produce biased outcomes. Who is responsible when an algorithm unfairly disadvantages a particular group? Is it the developers or the organizations implementing the technology? The absence of clear accountability mechanisms can create a culture of impunity, where biases are overlooked or inadequately addressed.

3. **Solutions for Reducing Bias:** Addressing bias in AI requires a multifaceted approach. Developers should prioritize diversity in their teams to bring varied perspectives to the design process. Additionally, rigorous testing and validation should be conducted to identify and rectify biases before AI systems are deployed.

The Ethics of Machine Decision-Making

1. **Autonomous Systems:** As AI technologies become more autonomous, ethical dilemmas around decision-making become increasingly complex. Consider self-driving cars, which must make split-second decisions in emergencies. How should a vehicle prioritize the safety of its passengers versus pedestrians in a potential accident scenario? Such moral dilemmas, often framed as "trolley problems," underscore the need for ethical frameworks guiding the programming of autonomous systems.

2. **Human Oversight:** Delegating decision-making to AI systems raises questions about the necessity of human oversight. While automation can enhance efficiency, it is crucial to ensure that humans remain involved, especially in high-stakes situations. Establishing clear guidelines for when and how human intervention is required can help maintain accountability and uphold ethical standards.

3. **Transparency in AI Decisions:** Transparency in AI decision-making processes is another ethical consideration. Many AI algorithms operate as "black boxes," making it difficult to understand how decisions are made. Ensuring transparency allows users to comprehend the rationale behind AI-generated outcomes, fostering trust and accountability.

Privacy and Surveillance Concerns

1. **Data Privacy:** The widespread use of AI often entails collecting and analyzing vast amounts of personal data, raising significant privacy concerns. Individuals may be unaware of how their data is being used or shared. Organizations must prioritize data privacy and obtain informed consent from users before collecting their information.

2. **Surveillance Technologies:** AI-powered surveillance technologies, such as facial recognition systems, have ignited debates about civil liberties and privacy rights. The potential for misuse by governments or corporations poses a threat to personal freedoms. Striking a balance between

reflects societal biases, the resulting algorithms can perpetuate and even amplify these biases. For instance, facial recognition technologies often perform less accurately for individuals with darker skin tones, raising concerns about discrimination in law enforcement and hiring practices.

2. **Accountability for Bias:** The issue of accountability arises when AI systems produce biased outcomes. Who is responsible when an algorithm unfairly disadvantages a particular group? Is it the developers or the organizations implementing the technology? The absence of clear accountability mechanisms can create a culture of impunity, where biases are overlooked or inadequately addressed.

3. **Solutions for Reducing Bias:** Addressing bias in AI requires a multifaceted approach. Developers should prioritize diversity in their teams to bring varied perspectives to the design process. Additionally, rigorous testing and validation should be conducted to identify and rectify biases before AI systems are deployed.

The Ethics of Machine Decision-Making

1. **Autonomous Systems:** As AI technologies become more autonomous, ethical dilemmas around decision-making become increasingly complex. Consider self-driving cars, which must make split-second decisions in emergencies. How should a vehicle prioritize the safety of its passengers versus pedestrians in a potential accident scenario? Such moral dilemmas, often framed as "trolley problems," underscore the need for ethical frameworks guiding the programming of autonomous systems.

2. **Human Oversight:** Delegating decision-making to AI systems raises questions about the necessity of human oversight. While automation can enhance efficiency, it is crucial to ensure that humans remain involved, especially in high-stakes situations. Establishing clear guidelines for when and how human intervention is required can help maintain accountability and uphold ethical standards.

3. **Transparency in AI Decisions:** Transparency in AI decision-making processes is another ethical consideration. Many AI algorithms operate as "black boxes," making it difficult to understand how decisions are made. Ensuring transparency allows users to comprehend the rationale behind AI-generated outcomes, fostering trust and accountability.

Privacy and Surveillance Concerns

1. **Data Privacy:** The widespread use of AI often entails collecting and analyzing vast amounts of personal data, raising significant privacy concerns. Individuals may be unaware of how their data is being used or shared. Organizations must prioritize data privacy and obtain informed consent from users before collecting their information.

2. **Surveillance Technologies:** AI-powered surveillance technologies, such as facial recognition systems, have ignited debates about civil liberties and privacy rights. The potential for misuse by governments or corporations poses a threat to personal freedoms. Striking a balance between

security and privacy is essential in developing ethical guidelines for these technologies.

The Need for Ethical Frameworks

1. **Establishing Ethical Guidelines:** To navigate the ethical dilemmas posed by AI, society must establish comprehensive ethical guidelines governing the development and implementation of these technologies. These frameworks should address issues of bias, accountability, privacy, and human oversight.

2. **Collaboration Across Sectors:** Addressing ethical concerns in AI requires collaboration among various stakeholders, including technologists, ethicists, policymakers, and the public. Fostering interdisciplinary dialogue can help develop solutions that balance innovation with ethical considerations.

3. **Public Engagement:** Engaging the public in discussions about AI ethics is crucial for fostering awareness and understanding. Individuals should have a voice in shaping policies and practices governing AI technologies, ensuring that their concerns and values are represented.

Conclusion: Navigating the Ethical Landscape

As AI continues to evolve and integrate into society, addressing the ethical dilemmas it presents is paramount. By prioritizing transparency, accountability, and human oversight, we can harness the potential of AI while mitigating its risks.

The ethical considerations surrounding AI technologies require thoughtful reflection and proactive measures. By fostering collaboration, establishing guidelines, and engaging the public, we can navigate the complex ethical landscape of AI, ensuring that technological advancements serve the greater good while upholding our shared values.

CHAPTER 15: THE IMPACT OF AI ON SOCIAL STRUCTURES

Introduction: The Social Landscape of AI
As artificial intelligence increasingly permeates everyday life, its impact on social structures becomes profound. AI technologies are reshaping how we work, communicate, and interact, creating both opportunities and challenges. This chapter examines the multifaceted effects of AI on employment, social relationships, and the broader fabric of society, highlighting the potential benefits and risks that accompany these transformations.

Transformation of Employment

1. **Job Displacement and Creation:** One of the most significant impacts of AI is its effect on employment. Automation could displace millions

of workers in sectors like manufacturing, transportation, and customer service. While some jobs will be lost, AI also has the potential to create new roles, particularly in technology and data analysis. The challenge lies in ensuring that displaced workers can transition to new opportunities without enduring prolonged unemployment or economic hardship.

2. **The Skills Gap:** The rise of AI has underscored a growing skills gap in the workforce. As industries adapt to new technologies, there is an increasing demand for workers with technical skills such as programming and data analysis. However, many individuals lack access to the education and training necessary to acquire these skills. Addressing this gap will require investment in education and retraining programs to equip workers with the tools they need to thrive in an AI-driven economy.

3. **Changing Work Dynamics:** AI is not only transforming the types of jobs available but also altering workplace dynamics. Remote work and AI-powered collaboration tools have enabled greater flexibility and connectivity, allowing teams to collaborate across geographical boundaries. However, this shift raises questions about work-life balance, employee engagement, and the potential for increased surveillance and control over workers.

Social Relationships in the Age of AI

1. **Interpersonal Communication:** The integration of AI into communication platforms has changed how people interact. From chatbots assisting with customer service to AI algorithms curating social media feeds, technology mediates our relationships

in new ways. While these tools can enhance connectivity, they may also contribute to a decline in face-to-face interactions and the quality of human relationships.

2. **Impact on Mental Health:** The rise of AI and technology in social interactions raises concerns about mental health and well-being. Studies indicate that excessive use of social media and technology can lead to feelings of isolation and anxiety. As AI becomes more prevalent in communication, it is essential to consider its potential effects on mental health and promote a balanced approach to technology use.

3. **Changing Definitions of Community:** AI technologies are reshaping the concept of community, enabling individuals to connect with others who share similar interests, regardless of location. Online communities facilitated by AI can foster collaboration and support but can also create echo chambers where dissenting opinions are silenced. The challenge lies in maintaining diverse and inclusive communities while navigating the influence of AI algorithms.

The Ethical Implications of AI on Society

1. **Surveillance and Privacy:** The increased use of AI for surveillance and monitoring raises ethical concerns about privacy and civil liberties. From facial recognition technologies to data collection practices, the potential for misuse of personal information is significant. Striking a balance between security and individual rights is crucial in navigating AI's ethical landscape.

2. **Social Inequality:** As AI technologies advance, there

is a risk of exacerbating existing social inequalities. Access to AI tools and education is often unequal, creating a divide between those who can harness the benefits of technology and those who cannot. Addressing this disparity will require targeted efforts to ensure equitable access to education, resources, and opportunities.

3. **Algorithmic Bias:** The algorithms powering AI systems can reflect and amplify societal biases. When used in decision-making processes—such as hiring or criminal justice—biased algorithms can perpetuate discrimination and inequality. Ensuring fairness and accountability in AI systems is essential for promoting social justice and protecting vulnerable populations.

Conclusion: Navigating the Social Impacts of AI

As AI continues to reshape social structures, it is essential to critically examine its implications for society. While AI presents opportunities for innovation and connectivity, it also poses challenges that must be addressed to ensure a fair and equitable future.

By fostering dialogue about the social impacts of AI and implementing policies that prioritize inclusivity, transparency, and ethical considerations, we can navigate the complexities of an AI-driven world. The path forward requires a collective commitment to harnessing AI's potential while safeguarding the values that underpin a just and equitable society.

CHAPTER 16: THE FUTURE OF AI AND HUMAN COEXISTENCE

Introduction: Envisioning a Collaborative Future
As artificial intelligence becomes more ingrained in daily life, the question of how humans and AI will coexist becomes increasingly pressing. The future landscape of our society will be shaped by our choices today, determining whether AI serves as a tool for enhancement or becomes a source of conflict. This chapter delves into the possibilities for harmonious coexistence, examining how we can leverage AI's strengths while addressing its challenges and ethical concerns.

The Potential for Collaboration

1. **Augmented Intelligence:** A promising vision for the future is the concept of augmented intelligence, where AI enhances human capabilities rather than

replacing them. By leveraging AI for data analysis, decision-making, and creative tasks, individuals can achieve greater efficiency and innovation. This partnership allows humans to focus on complex problem-solving and emotional intelligence—skills that AI cannot replicate. For example, in medicine, AI can assist doctors in diagnosing diseases more quickly and accurately, enabling them to provide better care for patients.

2. **Co-Creation with AI:** The potential for co-creation between humans and AI opens new avenues for artistic and scientific endeavors. Musicians, writers, and visual artists are already using AI tools to generate ideas, collaborate on projects, and push the boundaries of creativity. In scientific research, AI can analyze vast amounts of data to identify patterns and suggest new hypotheses, accelerating the pace of discovery. By embracing AI as a collaborative partner, we can unlock new forms of innovation that were previously unimaginable.

3. **Education and Lifelong Learning:** Preparing future generations for an AI-driven world necessitates a shift in education. Emphasizing skills such as critical thinking, creativity, and emotional intelligence will enable individuals to adapt to technological changes and work alongside AI systems. Lifelong learning will become essential as industries evolve, ensuring that people remain relevant and capable in an increasingly automated landscape.

Addressing Ethical and Regulatory Challenges

1. **Establishing Ethical Guidelines:** As we integrate AI into society, establishing ethical guidelines

is paramount. These guidelines should prioritize transparency, accountability, and fairness, ensuring that AI systems are designed and implemented responsibly. Collaboration among technologists, ethicists, policymakers, and the public is essential to create a framework that promotes ethical AI development and usage.

2. **Regulation and Oversight:** Governments and regulatory bodies will play a crucial role in shaping the future of AI. Appropriate regulations must be enacted to address issues such as data privacy, algorithmic bias, and accountability for AI-driven decisions. Striking the right balance between innovation and regulation will be key to fostering a safe and beneficial AI landscape.

3. **Promoting Public Awareness and Engagement:** Educating the public about AI, its benefits, and its risks is vital for fostering informed discussions about its societal role. Engaging citizens in conversations about AI's ethical implications and potential impacts will empower individuals to advocate for policies that align with their values and priorities.

A Vision for the Future

1. **Sustainable Development:** The future of AI can align with the principles of sustainable development. By harnessing AI's capabilities to tackle pressing global challenges such as climate change and resource management, we can work toward a more sustainable and equitable world. AI can optimize energy usage, improve agricultural practices, and facilitate the development of smart cities that reduce waste and enhance quality of life.

2. **Empowering Communities:** AI has the potential to empower marginalized communities by providing access to information, resources, and opportunities that were previously out of reach. By focusing on inclusive AI initiatives, we can bridge the digital divide and ensure that the benefits of technology are shared broadly, fostering a more equitable society.

3. **Human-Centric AI:** Ultimately, the future of AI must prioritize human values and well-being. Developing AI systems that respect and enhance human dignity, autonomy, and social connection will be essential for creating harmonious coexistence. By placing humans at the center of AI development, we can ensure that technology serves as a force for good.

Conclusion: Embracing the Future Together

The future of AI and human coexistence is filled with possibilities but requires careful consideration and proactive measures. By fostering collaboration, addressing ethical challenges, and prioritizing human values, we can shape a future where AI enhances our lives without undermining our humanity.

As we navigate this journey, it is imperative to embrace AI's potential while remaining vigilant about its risks. The choices we make today will determine the trajectory of our society, paving the way for a future where humans and AI coexist in a productive, ethical, and harmonious manner.

CHAPTER 17: THE ROLE OF AI IN SHAPING GLOBAL GOVERNANCE

Introduction: A New Era of Decision-Making

As artificial intelligence continues to evolve, its influence on global governance is becoming increasingly significant. AI technologies have the potential to transform how governments operate, enhance decision-making processes, and improve the delivery of public services. However, this transformation also raises critical questions about accountability, transparency, and the ethical implications of relying on AI in governance. This chapter explores the opportunities and challenges presented by AI in shaping global governance.

Enhancing Decision-Making and Policy Formulation

1. **Data-Driven Insights**: AI can analyze vast amounts

of data to identify trends, correlations, and insights that would be impossible for humans to discern. Governments can leverage AI to make data-driven decisions, enabling them to respond more effectively to societal needs. For instance, AI-powered predictive analytics can help governments anticipate public health crises, optimize resource allocation, and enhance disaster response strategies.

2. **Policy Simulation and Impact Assessment**: AI can simulate the potential outcomes of different policy choices, allowing policymakers to evaluate the impact of their decisions before implementation. By modeling complex systems and scenarios, AI can provide insights into the potential consequences of policies on various demographics, thereby facilitating more informed and equitable decision-making.

3. **Public Engagement and Feedback**: AI can enhance citizen engagement in the governance process by analyzing public sentiment and feedback. Natural language processing (NLP) technologies can sift through social media, forums, and surveys to gauge public opinion on specific issues, enabling governments to better understand the concerns and priorities of their constituents.

Challenges and Risks in AI Governance

1. **Bias and Inequity**: AI systems are not immune to biases inherent in the data they are trained on. If not carefully managed, AI can exacerbate existing inequalities in society by producing biased outcomes in decision-making processes. For example, biased algorithms in criminal justice or welfare distribution could disproportionately affect marginalized communities. Ensuring fairness and

equity in AI governance requires rigorous oversight and continuous evaluation of AI systems.

2. **Lack of Transparency**: The opaque nature of many AI algorithms can create challenges in accountability. Citizens may have difficulty understanding how decisions are made, leading to distrust in government processes. Transparent AI systems, where the decision-making process is clearly communicated, are essential for building public trust and ensuring accountability in governance.

3. **Ethical Dilemmas**: The use of AI in governance raises ethical concerns regarding privacy, surveillance, and autonomy. Governments must carefully balance the benefits of AI-driven insights with the potential risks of infringing on individual rights. Ethical frameworks must be developed to guide the responsible use of AI in public policy.

Global Collaboration and Governance Frameworks

1. **International Cooperation**: The global nature of AI technologies necessitates international collaboration to establish governance frameworks that address cross-border challenges. Countries must work together to develop standards for AI ethics, data protection, and accountability, ensuring that AI serves the interests of humanity as a whole.

2. **Multi-Stakeholder Approaches**: Engaging diverse stakeholders, including governments, businesses, civil society, and academia, is crucial for shaping AI governance. Multi-stakeholder collaboration fosters a more inclusive and comprehensive approach, allowing for the consideration of various

perspectives and interests in the development of AI policies.

3. **Regulatory Innovation**: As AI continues to evolve, traditional regulatory frameworks may need to adapt. Governments must explore innovative approaches to regulation that can keep pace with technological advancements while ensuring public safety and ethical standards. Agile regulatory frameworks can promote innovation while safeguarding the public interest.

Conclusion: Shaping a Responsible Future

The integration of AI into global governance presents both exciting opportunities and formidable challenges. By harnessing the potential of AI to enhance decision-making, promote transparency, and engage citizens, governments can create more effective and responsive governance structures. However, this transformation must be approached with caution, addressing the ethical implications and ensuring that AI systems are equitable, transparent, and accountable.

As we move forward, it is essential to prioritize the development of ethical AI governance frameworks that uphold human rights and promote social equity. The future of AI in governance depends on our ability to navigate these complexities, ensuring that technology serves as a force for good in shaping a more just and equitable world.

CHAPTER 18: THE FUTURE OF AI AND HUMAN SOCIETY

Introduction: A New Frontier

As we stand on the brink of an era defined by artificial intelligence, the future of human society is poised for transformation. AI has the potential to revolutionize nearly every aspect of our lives, from the way we work and communicate to how we govern and interact with one another. However, this unprecedented change also brings profound questions about our identity, values, and the societal structures that will emerge in response to AI's growing presence. This chapter delves into the possibilities and challenges that lie ahead in a future shaped by AI.

Envisioning a Collaborative Future

1. **Human-AI Collaboration**: Rather than viewing AI as a threat, we can imagine a future where humans

and machines collaborate to enhance productivity and creativity. In various fields, such as healthcare, education, and engineering, AI can complement human skills by providing insights, automating repetitive tasks, and assisting in complex problem-solving. This partnership can lead to improved outcomes and innovation, allowing humans to focus on what they do best—creativity, empathy, and critical thinking.

2. **Augmented Human Capabilities**: The integration of AI into everyday life may lead to the augmentation of human capabilities. Technologies like brain-computer interfaces could enhance cognitive functions, allowing individuals to process information more rapidly and make better-informed decisions. Such advancements can empower people to tackle challenges that require deep analytical skills, ultimately enriching the human experience and fostering a more knowledgeable society.

3. **A More Equitable Society**: AI has the potential to promote social equity by improving access to essential services, such as education and healthcare. AI-driven personalized learning platforms can provide tailored educational experiences that address individual learning needs, while telemedicine powered by AI can enhance access to healthcare in underserved communities. If deployed thoughtfully, AI can help bridge existing gaps in society and create a more inclusive future.

Navigating the Challenges Ahead

1. **Ethical Considerations**: As AI becomes more integrated into our lives, ethical considerations must take center stage. The development of AI should prioritize transparency, fairness, and

accountability to ensure that it aligns with societal values. The ethical implications of AI must be actively addressed through robust regulatory frameworks and ongoing public dialogue, ensuring that technological advancements do not come at the expense of human rights.

2. **Job Displacement and Economic Transition**: The rise of AI and automation raises concerns about job displacement and economic inequality. While AI can enhance productivity, it may also render certain jobs obsolete. It is crucial to implement proactive measures, such as retraining programs and universal basic income, to support workers affected by these changes. By prioritizing the human workforce alongside technological advancement, we can create a balanced economy that empowers individuals rather than marginalizing them.

3. **Surveillance and Privacy**: The integration of AI into daily life also presents significant challenges related to surveillance and privacy. As AI systems collect and analyze vast amounts of data, there is a risk that individuals' privacy rights may be compromised. Society must engage in discussions about data ownership, consent, and surveillance practices to ensure that technology serves the public interest without infringing on personal freedoms.

A Call to Action: Shaping Our Future

1. **Inclusive Governance**: As AI technology advances, inclusive governance frameworks must be established to involve diverse stakeholders in decision-making processes. Policymakers, technologists, ethicists, and citizens must collaborate to develop regulations that govern AI development and deployment. This collaborative approach ensures that the voices of all affected

communities are heard and considered, leading to policies that reflect the collective values of society.

2. **Education and Awareness**: Fostering an informed society is essential in navigating the complexities of an AI-driven future. Education should include discussions about AI's potential and its implications, equipping individuals with the knowledge and skills needed to thrive in a technology-rich environment. By promoting awareness of AI's capabilities and limitations, we can encourage responsible engagement with technology.

3. **Embracing Humanity**: Finally, as we embrace the advancements of AI, it is essential to remember the core values that define us as humans—compassion, empathy, and creativity. The future of AI should not only focus on technological innovation but also on enhancing the human experience and fostering connections among individuals. By prioritizing our shared humanity, we can ensure that AI serves to uplift and empower rather than isolate and alienate.

Conclusion: A Shared Responsibility

The future of AI and human society is a shared responsibility that demands thoughtful consideration, proactive measures, and a commitment to ethical principles. As we navigate this uncharted territory, we must embrace the opportunities AI presents while remaining vigilant about its potential risks. By fostering collaboration, inclusivity, and transparency, we can build a future where AI enhances our lives and enriches society as a whole.

CHAPTER 19: THE SECRETS OF AI AND ROBOTICS

Introduction: Unveiling the Enigma

Artificial Intelligence (AI) and robotics are among the most transformative technologies of our time, revolutionizing industries, redefining the workplace, and changing the way we interact with the world. While much has been said about their capabilities and potential, the inner workings of AI and robotics often remain shrouded in mystery. This chapter aims to peel back the layers, uncovering the secrets that define these technologies and their implications for humanity.

The Building Blocks of AI

1. **Machine Learning**: At the core of modern AI is machine learning, a subset of AI that enables machines to learn from data and improve their performance over time without explicit

programming. Machine learning algorithms analyze vast datasets to identify patterns, make predictions, and inform decisions. Understanding the principles of supervised, unsupervised, and reinforcement learning is essential to grasp how AI systems function and adapt.

2. **Neural Networks**: Inspired by the human brain, neural networks are computational models that mimic the way neurons work together to process information. These networks consist of layers of interconnected nodes that process data in a hierarchical manner. Deep learning, a branch of machine learning that utilizes neural networks with multiple layers, has propelled AI to new heights, enabling breakthroughs in image and speech recognition, natural language processing, and more.

3. **Natural Language Processing (NLP)**: NLP is the AI discipline focused on enabling machines to understand and generate human language. By analyzing text and speech data, NLP algorithms can interpret context, sentiment, and intent, allowing AI systems to engage in human-like conversations and automate tasks such as customer service and content generation.

Robotics: The Physical Manifestation of AI

1. **Sensors and Perception**: Robotics relies heavily on sensors to perceive the environment. These sensors gather data about surroundings, enabling robots to navigate, identify objects, and interact with the world. Understanding the role of vision systems, tactile sensors, and environmental sensors is crucial for appreciating how robots interpret and respond to their environment.

2. **Actuation and Movement**: Actuators are the mechanisms that enable robots to move and manipulate objects. From electric motors to hydraulic systems, the choice of actuation technology significantly impacts a robot's capabilities and performance. Exploring the intricacies of robotic movement sheds light on how robots can achieve complex tasks, from assembly line work to delicate surgeries.

3. **Autonomy and Decision-Making**: One of the most intriguing aspects of robotics is the ability to make autonomous decisions. Through algorithms that process sensory data and evaluate potential actions, robots can operate independently in dynamic environments. The balance between autonomy and human oversight is critical, raising questions about the ethical implications of allowing robots to make decisions without human intervention.

The Ethical Secrets of AI and Robotics

1. **Bias and Fairness**: AI systems are not immune to biases present in the data used to train them. Understanding how biases can lead to unfair outcomes in decision-making is essential for developing responsible AI. Awareness of data quality, representation, and the importance of diverse datasets is crucial in addressing bias in AI applications.

2. **Accountability and Transparency**: As AI systems increasingly influence critical areas such as healthcare, law enforcement, and finance, questions of accountability and transparency arise. Who is responsible when an AI makes a mistake? Establishing clear lines of accountability and ensuring transparency in AI decision-making

processes are essential for building trust in these technologies.

3. **The Future of Work**: The rise of AI and robotics is poised to disrupt traditional job markets. While automation offers the potential for increased efficiency and productivity, it also raises concerns about job displacement and economic inequality. Understanding the implications of these technologies for the workforce is essential for preparing society for the changes ahead.

Preparing for an AI-Driven Future

1. **Education and Reskilling**: As AI and robotics continue to evolve, the demand for skilled workers in technology and data science will grow. Educational institutions and organizations must adapt to provide relevant training and reskilling programs, preparing individuals for careers in an AI-driven economy.

2. **Public Engagement and Awareness**: Engaging the public in discussions about AI and robotics is crucial for fostering understanding and informed decision-making. Raising awareness of the capabilities, limitations, and ethical considerations of these technologies will empower individuals to navigate the changes they bring.

3. **Regulatory Frameworks**: Developing comprehensive regulatory frameworks is vital to ensure that AI and robotics are deployed responsibly and ethically. Policymakers must work with technologists, ethicists, and the public to create guidelines that balance innovation with

societal needs and values.

Conclusion: Embracing the Future Together

The secrets of AI and robotics are not just technical intricacies but profound questions about humanity's relationship with technology. As we continue to explore the potential of these technologies, it is essential to embrace collaboration, transparency, and ethical considerations. By doing so, we can harness the power of AI and robotics to create a future that benefits all of humanity, ensuring that technology serves as a tool for empowerment rather than a source of division.

CHAPTER 20: THE FUTURE OF AI AND ROBOTICS: OPPORTUNITIES AND THREATS

Introduction: Navigating the Horizon

As we stand on the brink of an era defined by rapid technological advancement, the future of artificial intelligence (AI) and robotics promises both remarkable opportunities and daunting threats. The capabilities of AI are evolving at an unprecedented pace, affecting every aspect of our lives, from healthcare and education to transportation and entertainment. Understanding what lies ahead requires a balanced perspective that acknowledges the benefits while being wary of the potential risks.

Transformative Opportunities

1. **Revolutionizing Healthcare**: AI has the potential to transform healthcare delivery by improving diagnostics, personalizing treatment, and enhancing patient care. Advanced algorithms can analyze medical images with precision, enabling earlier detection of diseases like cancer. Robotics can assist in surgeries, making procedures less invasive and reducing recovery times. As AI systems continue to learn from vast datasets, their ability to provide accurate predictions and insights will revolutionize how we approach healthcare.

2. **Enhancing Education**: The integration of AI in education can create personalized learning experiences tailored to individual students' needs. Adaptive learning platforms can assess a student's strengths and weaknesses, providing targeted resources to help them succeed. Robotics in classrooms can facilitate hands-on learning, engaging students in STEM subjects through interactive experiences. By harnessing these technologies, we can improve educational outcomes and make learning more accessible.

3. **Sustainable Solutions**: AI and robotics can play a pivotal role in addressing pressing environmental challenges. Smart grids can optimize energy distribution, reducing waste and promoting the use of renewable resources. Autonomous drones can monitor wildlife populations and assist in conservation efforts. By leveraging technology to create sustainable practices, we can mitigate climate change and preserve our planet for future generations.

4. **Workforce Transformation**: While automation

raises concerns about job displacement, it also presents opportunities for workforce transformation. By automating repetitive and dangerous tasks, AI and robotics can free human workers to engage in more creative, strategic, and fulfilling roles. The rise of AI may lead to the creation of new industries and job categories that we cannot yet imagine, ultimately reshaping the nature of work.

The Looming Threats

1. **Job Displacement and Economic Inequality**: The automation of jobs poses a significant threat to the workforce, particularly for low-skilled laborers. As AI systems become capable of performing tasks traditionally done by humans, there is a real risk of widespread job loss, leading to economic instability and increased inequality. Society must proactively address these challenges by investing in retraining programs and supporting workers transitioning to new roles.

2. **Ethical and Privacy Concerns**: The increasing use of AI raises ethical dilemmas regarding privacy, surveillance, and data security. As AI systems become more integrated into daily life, concerns about the collection and use of personal data grow. The potential for AI to be used in surveillance and social control raises questions about civil liberties and human rights. It is crucial to establish guidelines that protect individual privacy and ensure the ethical use of AI technologies.

3. **Autonomous Weapons and Warfare**: The development of AI-driven autonomous weapons systems presents profound ethical and security challenges. The potential for machines to make life-and-death decisions raises questions about

accountability and the ethics of warfare. As nations race to develop these technologies, the risk of escalation and conflict increases. International cooperation and regulation are essential to prevent a new arms race driven by AI.

4. **Existential Risks**: The most profound threat posed by advanced AI lies in its potential to surpass human intelligence. If we create superintelligent AI without proper safeguards, we risk losing control over these systems. The development of AI that can outthink and outmaneuver human decision-making could lead to unintended consequences that threaten humanity's existence. Engaging in robust discussions about the implications of advanced AI is critical to ensure its alignment with human values.

Navigating the Future

1. **Fostering Collaboration**: The future of AI and robotics demands collaboration among technologists, policymakers, ethicists, and the public. By engaging diverse stakeholders in discussions about technology's impact, we can create frameworks that guide responsible innovation and ensure that the benefits are shared equitably.

2. **Implementing Ethical Guidelines**: Establishing ethical guidelines for AI and robotics is essential to navigate the challenges ahead. Organizations, governments, and industry leaders must work together to create standards that prioritize transparency, accountability, and human rights.

3. **Investing in Education and Training**: Preparing the workforce for an AI-driven future requires

significant investment in education and training. By equipping individuals with the skills necessary to thrive in an evolving job market, we can mitigate the impacts of automation and ensure that workers can adapt to new roles.

4. **Promoting Public Awareness**: Raising public awareness about the implications of AI and robotics is crucial for informed decision-making. By fostering a dialogue around these technologies, we can empower individuals to engage with the changes that are shaping their lives and advocate for policies that align with their values.

Conclusion: A Collective Responsibility

The future of AI and robotics is a landscape filled with both remarkable opportunities and formidable challenges. As we move forward, it is essential to embrace the transformative potential of these technologies while remaining vigilant about the ethical and social implications they present. By fostering collaboration, implementing ethical guidelines, and promoting public awareness, we can navigate the complexities of an AI-driven world and work toward a future that benefits all of humanity.

CHAPTER 21: THE HUMAN ELEMENT IN AI AND ROBOTICS

Introduction: The Heart of Technology

As we venture further into the realm of artificial intelligence (AI) and robotics, it's essential to remember that these technologies are designed to serve humanity. Despite their growing complexity and capabilities, the fundamental purpose of AI and robotics should be to enhance human life, not replace it. This chapter explores the critical role of the human element in the development, implementation, and interaction with AI and robotics, emphasizing the need for empathy, ethics, and human-centered design.

The Human-Centric Approach

1. **Empathy in Design**: The development of AI and robotics must prioritize the needs and experiences of users. By incorporating empathy into the design

process, developers can create technologies that resonate with people, understanding their pain points and desires. This human-centric approach fosters trust and ensures that technology enhances the user experience rather than complicates it.

2. **Collaboration Between Humans and Machines**: The most effective applications of AI and robotics involve collaboration rather than replacement. By leveraging the strengths of both humans and machines, we can achieve remarkable results. For instance, in healthcare, AI can assist doctors in diagnosing diseases more accurately, while human professionals provide the empathy and emotional support that machines cannot replicate.

3. **Maintaining Human Oversight**: As AI systems become more autonomous, maintaining human oversight is crucial. Humans must remain in control of decision-making processes, especially in high-stakes environments such as healthcare, finance, and law enforcement. By establishing clear lines of accountability and oversight, we can mitigate risks associated with automated decision-making.

The Ethical Imperative

1. **Values and Ethics in AI Development**: As AI and robotics evolve, developers must consider the ethical implications of their technologies. What values do these systems reflect? How do they align with human rights and social justice? Engaging ethicists, sociologists, and diverse communities in the development process ensures that technologies uphold ethical standards and promote equity.

2. **Addressing Bias and Discrimination**: AI systems can inadvertently perpetuate bias if not carefully designed. Developers must actively work to identify

and eliminate biases in algorithms, ensuring that AI applications do not discriminate based on race, gender, or socioeconomic status. Incorporating diverse perspectives in the design process is essential for creating fair and just systems.

3. **Transparency and Accountability**: Users must understand how AI systems operate and the data they utilize. Promoting transparency in AI development helps build trust and allows users to hold developers accountable for their technologies. Organizations should provide clear explanations of how algorithms make decisions and the potential consequences of those decisions.

The Future of Work: Redefining Roles

1. **Upskilling and Reskilling**: As AI and robotics reshape industries, the workforce must adapt. Investing in upskilling and reskilling programs equips workers with the necessary skills to thrive in an AI-driven economy. By focusing on developing critical thinking, creativity, and interpersonal skills, workers can complement automated systems and contribute to a more productive workforce.

2. **Emphasizing Soft Skills**: While technical skills are essential, soft skills such as communication, empathy, and collaboration will become increasingly valuable in the workplace. As machines handle repetitive tasks, human workers can focus on roles that require emotional intelligence and interpersonal interactions. Organizations must prioritize cultivating these skills in their workforce.

3. **Fostering a Culture of Innovation**: Encouraging a culture of innovation within organizations empowers employees to experiment with new ideas and embrace technological advancements. By

fostering a supportive environment that values creativity and collaboration, companies can harness the potential of AI and robotics while prioritizing the human element.

Building Trust in AI and Robotics

1. **User Education**: Educating users about AI and robotics is crucial for building trust. By providing information about how these technologies work, their limitations, and their potential risks, organizations can empower users to engage with technology confidently. Training programs and resources can help demystify AI and robotics, fostering a better understanding of their applications.

2. **Promoting Community Engagement**: Involving communities in discussions about AI and robotics ensures that diverse perspectives are considered. Public forums, workshops, and collaborations with local organizations can facilitate dialogue about the ethical implications and societal impact of these technologies. Engaging communities helps build trust and fosters a sense of ownership in the development process.

3. **Establishing Regulatory Frameworks**: Governments and regulatory bodies play a vital role in ensuring the ethical use of AI and robotics. By developing regulations that prioritize transparency, accountability, and user rights, societies can create a framework that fosters trust in these technologies. Collaboration between industry, academia, and government is essential for establishing effective

policies that protect individuals and communities.

Conclusion: The Future is Human

As AI and robotics continue to evolve, the human element must remain at the forefront of innovation. By prioritizing empathy, ethics, and collaboration, we can ensure that these technologies serve to enhance rather than diminish the human experience. The future of AI and robotics should not only be about technological advancement but also about creating a society where technology empowers individuals and communities.

CHAPTER 22: THE DANGERS OF AUTONOMOUS AI AND ROBOTS

Introduction: The Rise of Autonomy

As we stand on the brink of an era dominated by autonomous AI and robots, we must confront the potential dangers that accompany this technological revolution. While the promise of increased efficiency and productivity is alluring, the implications of allowing machines to operate independently raise significant concerns. This chapter delves into the various dangers posed by autonomous AI and robots, exploring the ethical, social, and safety challenges that must be addressed as we integrate these systems into our daily lives.

The Ethical Quagmire

    1. **Decision-Making Without Accountability**: One of

the most pressing ethical concerns surrounding autonomous AI is the issue of accountability. When machines make decisions—especially in high-stakes situations—who is responsible for the outcomes? If an autonomous vehicle is involved in an accident, should the blame fall on the manufacturer, the software developers, or the user? This ambiguity complicates our understanding of responsibility in an increasingly automated world.

2. **Moral Dilemmas in Programming**: The challenge of programming ethical decision-making into AI systems is daunting. Consider the "trolley problem," a classic ethical dilemma where one must choose between saving multiple lives at the expense of one. How should an autonomous vehicle be programmed to respond in such scenarios? The difficulty of encoding moral judgments into algorithms raises questions about the values we prioritize and the implications of those choices.

3. **Potential for Misuse**: Autonomous AI and robotics can be weaponized, leading to ethical quandaries that society must grapple with. Drones, autonomous weapons, and surveillance systems represent a dark side of technology that could be exploited for malicious purposes. Ensuring that these technologies are used ethically and responsibly is essential to prevent their misuse.

Social Implications

1. **Job Displacement and Economic Inequality**: As autonomous AI and robots become capable of performing tasks traditionally carried out by humans, the potential for widespread job displacement looms large. Industries such as transportation, manufacturing, and customer service are particularly vulnerable. Without

proactive measures to retrain workers, society risks exacerbating economic inequality and leaving many individuals without viable employment options.

2. **Loss of Human Interaction**: The increasing reliance on autonomous systems may lead to a decline in human-to-human interactions. As robots take over roles in customer service, healthcare, and even companionship, the fabric of social connections may fray. This loss of human interaction can contribute to feelings of isolation and disconnection, impacting mental health and well-being.

3. **Erosion of Skills**: The more we rely on autonomous systems, the more we risk losing essential skills. As technology takes over tasks previously performed by humans, there is a danger that individuals may become complacent and less adept at problem-solving, critical thinking, and manual tasks. This erosion of skills could have long-term implications for workforce adaptability and resilience.

Safety Concerns

1. **Malfunctioning Systems**: Autonomous AI and robots are not infallible. Software bugs, hardware failures, and unforeseen interactions can lead to malfunctions with potentially dangerous consequences. For instance, if an autonomous vehicle's sensors fail, it may misinterpret its surroundings, leading to accidents and endangering lives. Ensuring robust safety measures and fail-safes is crucial as we integrate these systems into our society.

2. **Cybersecurity Risks**: The interconnected nature of autonomous systems makes them vulnerable

to cyberattacks. Malicious actors could exploit vulnerabilities in AI algorithms or robotic systems, potentially leading to disastrous outcomes. For example, hacking into an autonomous vehicle could allow an attacker to take control of the car, endangering the occupants and others on the road. As we embrace autonomous technology, prioritizing cybersecurity becomes paramount.

3. **Unintended Consequences**: The complexities of AI and robotics can lead to unintended consequences that are difficult to predict. For example, the deployment of autonomous delivery drones may disrupt local ecosystems or pose risks to wildlife. Understanding the full scope of potential impacts is essential to mitigate risks and ensure responsible deployment.

Regulatory Challenges

1. **Lack of Regulatory Frameworks**: The rapid advancement of autonomous AI and robots has outpaced the development of regulatory frameworks. Governments and regulatory bodies are often ill-equipped to address the ethical, legal, and safety challenges posed by these technologies. Establishing comprehensive regulations is essential to ensure that autonomous systems are developed and deployed responsibly.

2. **Balancing Innovation with Oversight**: Striking the right balance between fostering innovation and ensuring adequate oversight is a delicate task. While excessive regulation can stifle technological advancement, insufficient oversight can lead to dangerous outcomes. Policymakers must navigate this tension to create an environment that encourages innovation while safeguarding public safety and ethical standards.

3. **Global Cooperation**: The challenges posed by autonomous AI and robots transcend national borders, necessitating global cooperation. Collaborative efforts among governments, industry leaders, and international organizations are essential to establish guidelines and standards that promote the safe and ethical use of these technologies worldwide.

Conclusion: A Cautious Path Forward

As we embrace the potential of autonomous AI and robots, we must remain vigilant about the dangers they pose. By prioritizing ethics, social responsibility, and safety, we can harness the benefits of these technologies while mitigating their risks. The future of AI and robotics should be guided by a commitment to enhancing human life, ensuring that these systems serve as tools for progress rather than sources of harm.

CHAPTER 23: THE FUTURE OF WORK IN AN AI-DRIVEN WORLD

Introduction: The Evolving Landscape of Employment

As AI and automation continue to advance, they are poised to transform the future of work in ways we are only beginning to understand. While these technologies promise increased efficiency and productivity, they also raise concerns about job displacement and the skills required to thrive in an increasingly automated economy. This chapter explores the implications of AI and robotics on the job market, highlighting both the opportunities and challenges that lie ahead.

The Changing Nature of Work

1. **Automation and Job Displacement**: One of the most immediate impacts of AI and automation is the potential for widespread job displacement. Many jobs that involve routine tasks, such as data

entry, assembly line work, and even some aspects of customer service, are at risk of being replaced by machines. While some industries may adapt by creating new roles, the transition may not be smooth for all workers, leading to significant disruptions in the job market.

2. **Emergence of New Roles**: Despite the threat of displacement, AI and automation also have the potential to create new jobs. As technology advances, there will be an increasing demand for skilled workers in fields such as AI development, robotics maintenance, and data analysis. Additionally, roles that require human creativity, emotional intelligence, and complex problem-solving are likely to become more valuable, as these are areas where machines still fall short.

3. **Reskilling and Upskilling**: To navigate the changing landscape of work, individuals will need to prioritize reskilling and upskilling. Lifelong learning will become essential as workers adapt to new technologies and seek to enhance their skill sets. Organizations must also invest in training programs to help employees transition to new roles and develop the competencies needed for the future.

The Gig Economy and Flexible Work Arrangements

1. **Rise of the Gig Economy**: The gig economy, characterized by short-term contracts and freelance work, has gained popularity in recent years. While this shift offers flexibility and autonomy for workers, it also raises questions about job security, benefits, and workers' rights. As more individuals turn to gig work, there will be a need for regulatory frameworks that protect these workers and ensure fair treatment.

2. **Remote Work Revolution**: The COVID-19 pandemic accelerated the adoption of remote work, prompting organizations to rethink traditional office environments. While remote work offers flexibility and can enhance productivity, it also presents challenges related to collaboration, team cohesion, and work-life balance. As companies embrace hybrid work models, finding effective ways to support employees will be crucial.

3. **Technology-Enabled Collaboration**: Advances in communication technology will continue to reshape how teams collaborate, both in-person and remotely. Tools such as video conferencing, project management software, and AI-driven collaboration platforms will facilitate teamwork across geographical boundaries. However, organizations must be mindful of the potential for digital fatigue and strive to foster a healthy work culture.

The Ethical Implications of AI in the Workplace

1. **Bias and Discrimination**: The use of AI in hiring and performance evaluation raises ethical concerns regarding bias and discrimination. If AI systems are trained on historical data that reflects existing biases, they may perpetuate inequalities in hiring and promotions. Organizations must implement measures to ensure that AI is used fairly and transparently in employment decisions.

2. **Surveillance and Privacy**: As companies adopt AI-driven monitoring tools to track employee performance and productivity, concerns about privacy and surveillance grow. Workers may feel that their every move is being monitored, leading

to anxiety and distrust. Striking a balance between productivity and employee privacy will be essential in maintaining a healthy workplace environment.

3. **Human-AI Collaboration**: The future of work will not only involve machines replacing human roles but also collaboration between humans and AI. This partnership has the potential to enhance productivity and creativity, but it requires a thoughtful approach to integrating AI into workflows. Organizations must foster a culture of collaboration, encouraging employees to embrace technology as a tool rather than a threat.

Conclusion: Preparing for an AI-Driven Future

As we look to the future, the impact of AI and automation on the workforce will be profound. By understanding the changing landscape of work and proactively addressing the challenges and opportunities it presents, individuals and organizations can position themselves for success. Embracing lifelong learning, fostering inclusive work environments, and ensuring ethical practices in the use of AI will be crucial as we navigate this new era.

CHAPTER 24: THE SECRETS OF AI: UNRAVELING THE TECHNOLOGY BEHIND THE CURTAIN

Introduction: Understanding the AI Landscape

As artificial intelligence (AI) continues to permeate our daily lives, it's crucial to understand the underlying technologies that drive this transformative force. This chapter delves into the secrets of AI, exploring how machine learning, neural networks, and data processing work together to create systems that can learn, adapt, and make decisions. Additionally, we will examine the ethical implications of these technologies and their potential impact on society.

The Foundations of AI

1. **What is Artificial Intelligence?** At its core, artificial intelligence refers to the simulation of human intelligence processes by machines, especially computer systems. These processes include learning (acquiring information and rules for using it), reasoning (using rules to reach approximate or definite conclusions), and self-correction. AI can be categorized into two main types: narrow AI, which is designed for specific tasks, and general AI, which aims to replicate human cognitive abilities across a wide range of activities.

2. **Machine Learning: Teaching Machines to Learn**: Machine learning is a subset of AI that involves training algorithms to recognize patterns and make decisions based on data. Instead of being explicitly programmed for specific tasks, machine learning models improve their performance through experience. For instance, a machine learning model can be trained to identify images by being exposed to thousands of labeled examples, gradually learning to distinguish between different objects.

3. **Deep Learning: The Power of Neural Networks**: Deep learning, a subset of machine learning, employs artificial neural networks to process and analyze vast amounts of data. Neural networks are inspired by the human brain's architecture, consisting of interconnected nodes (neurons) that work together to process information. Deep learning has revolutionized fields such as computer vision, natural language processing, and speech recognition, enabling machines to perform tasks that were once thought to be exclusive to humans.

4. **Data: The Fuel for AI**: The effectiveness of AI systems largely depends on the quality and quantity of data they are trained on. Big data, characterized by its volume, velocity, and variety, provides the raw material for machine learning algorithms. Organizations must invest in data collection, cleaning, and management to ensure that their AI systems operate effectively and responsibly.

The Ethical Considerations of AI

1. **Bias in AI**: One of the most pressing ethical concerns surrounding AI is bias. AI systems trained on historical data may inadvertently perpetuate existing prejudices and inequalities. For example, facial recognition technology has been shown to have higher error rates for people of color due to a lack of diversity in the training data. Addressing bias in AI is essential to ensuring that these technologies do not reinforce societal inequalities.

2. **Transparency and Accountability**: As AI systems become increasingly complex, understanding their decision-making processes can be challenging. The "black box" nature of many AI models raises questions about accountability, especially when decisions made by AI have significant consequences. Ensuring transparency in AI systems, including how they make decisions and the data used, is vital for building trust and ensuring ethical use.

3. **Privacy Concerns**: The data-driven nature of AI also raises concerns about privacy. Collecting and analyzing large amounts of personal data

can infringe on individuals' privacy rights. Organizations must implement robust data protection measures and adhere to ethical standards to safeguard user information and build public trust in AI technologies.

4. **The Future of Employment**: The rise of AI has implications for the workforce, as discussed in the previous chapter. As machines take on more tasks, workers may face displacement and the need to adapt to new roles. Organizations and governments must work together to provide retraining programs and support systems to help individuals transition to an AI-driven job market.

Conclusion: The Path Forward

As we unravel the secrets of AI, it becomes clear that this technology holds immense potential to transform our lives and society. However, with that potential comes the responsibility to address the ethical challenges that arise. By fostering transparency, addressing bias, and ensuring responsible data practices, we can harness the power of AI for the greater good. As we move forward into an increasingly automated world, understanding the technology behind AI will be essential for navigating the complexities of our future.

CHAPTER 25: THE RISE OF AUTONOMOUS SYSTEMS: OPPORTUNITIES AND CHALLENGES

Introduction: A New Era of Autonomy

As we stand on the brink of a new era defined by autonomous systems, it is essential to explore the implications of this technological revolution. From self-driving cars to drones, autonomous systems have the potential to reshape industries, enhance productivity, and improve our quality of life. However, alongside these opportunities come significant challenges, including safety concerns, ethical dilemmas, and the need for regulatory frameworks. This chapter will examine the rise of autonomous systems, highlighting both their transformative potential and the hurdles we must overcome.

The Growth of Autonomous Systems

1. **Understanding Autonomy**: At its core, autonomy refers to the ability of a system to operate independently without human intervention. Autonomous systems use advanced technologies, including AI, machine learning, and robotics, to perform tasks traditionally carried out by humans. These systems can analyze data, make decisions, and learn from their experiences, allowing them to navigate complex environments.

2. **Self-Driving Vehicles**: One of the most prominent examples of autonomous systems is self-driving cars. Companies like Tesla, Waymo, and others are at the forefront of developing vehicles equipped with AI-powered technology capable of navigating roads, avoiding obstacles, and making real-time decisions. While self-driving cars promise to enhance safety by reducing human error, they also raise questions about liability, insurance, and the future of transportation jobs.

3. **Drones and Robotics**: Autonomous drones and robots are being used in various sectors, from agriculture to logistics. Drones can monitor crops, deliver packages, and even assist in search and rescue operations. Similarly, robots are being deployed in warehouses and manufacturing facilities to streamline processes and reduce costs. As these technologies become more advanced, they offer the potential to increase efficiency and reduce operational risks.

The Opportunities of Autonomous Systems

1. **Enhanced Efficiency**: Autonomous systems can operate around the clock, minimizing downtime

and increasing productivity. For instance, self-driving trucks can transport goods without the need for breaks, reducing delivery times and operational costs. In agriculture, drones can monitor vast fields more quickly than human workers, enabling more efficient resource management.

2. **Improved Safety**: By eliminating human error, autonomous systems have the potential to enhance safety in various contexts. Self-driving cars can reduce the number of accidents caused by distracted or impaired drivers. In industrial settings, robots can take on dangerous tasks, reducing the risk of injury to human workers.

3. **Environmental Benefits**: Autonomous systems can also contribute to sustainability efforts. For example, optimized logistics powered by AI can reduce fuel consumption and lower greenhouse gas emissions. In agriculture, drones can help optimize pesticide usage, minimizing environmental impact while maximizing crop yields.

The Challenges Ahead

1. **Safety and Reliability**: While autonomous systems promise enhanced safety, their reliability is paramount. The technology must undergo rigorous testing to ensure it can handle complex real-world scenarios. High-profile accidents involving autonomous vehicles have raised concerns about the safety of these systems and the potential consequences of malfunctions.

2. **Ethical Dilemmas**: The rise of autonomous systems also brings ethical challenges. For instance, self-

driving cars must be programmed to make split-second decisions in emergency situations, raising questions about how to prioritize human lives. Should the vehicle protect its occupants at all costs, or should it consider the safety of pedestrians? These dilemmas highlight the need for ethical frameworks to guide the development of autonomous technologies.

3. **Job Displacement**: As autonomous systems become more prevalent, the potential for job displacement looms large. Workers in industries such as trucking, manufacturing, and agriculture may find their roles at risk as machines take over tasks previously performed by humans. Addressing this issue requires proactive measures, including retraining programs and support systems for affected workers.

4. **Regulatory and Legal Frameworks**: The rapid development of autonomous systems outpaces existing regulations. Policymakers must work to create frameworks that address liability, safety standards, and ethical considerations. Failure to establish clear guidelines could hinder the adoption of these technologies and create confusion in the marketplace.

Conclusion: Navigating the Future of Autonomy

The rise of autonomous systems presents both unprecedented opportunities and significant challenges. As we embrace this technological revolution, it is essential to address the ethical, social, and regulatory implications that accompany it. By fostering collaboration between technologists, policymakers, and society, we can navigate the complexities of autonomous systems and ensure that their benefits are realized while minimizing risks. The future of autonomy is bright, but it requires thoughtful stewardship to ensure it serves the greater good.

CHAPTER 26: THE ETHICAL IMPLICATIONS OF AI IN DAILY LIFE

Introduction: AI in Our Everyday Existence

Artificial Intelligence (AI) is no longer a futuristic concept; it has seamlessly integrated into our daily lives, influencing how we communicate, work, shop, and even make decisions. From personal assistants like Siri and Alexa to recommendation algorithms on streaming services, AI shapes our experiences in profound ways. However, with these advancements come ethical implications that must be scrutinized. This chapter delves into the ethical considerations surrounding AI's presence in our everyday lives, examining how it affects our privacy, autonomy, and societal norms.

The Ubiquity of AI

1. **Personal Assistants**: AI-powered personal assistants have become common in households, helping users manage their schedules, control smart home devices, and access information quickly. While these tools enhance convenience, they also raise concerns about data privacy and the extent to which these assistants monitor user behavior.

2. **Recommendation Algorithms**: Streaming platforms, online retailers, and social media networks utilize AI algorithms to curate content and recommendations based on user preferences. While this personalization improves user experience, it can create echo chambers, limit exposure to diverse viewpoints, and reinforce biases in content selection.

3. **Smart Home Devices**: From thermostats that learn your habits to security cameras equipped with facial recognition, smart home devices leverage AI to enhance comfort and security. However, these technologies also raise questions about surveillance and the collection of personal data, often without explicit user consent.

Ethical Considerations of AI Integration

1. **Privacy Concerns**: One of the most pressing ethical concerns surrounding AI in daily life is the potential for privacy violations. AI systems often collect vast amounts of data to function effectively, leading to concerns about how this data is stored, shared, and used. For example, personal assistants may record conversations to improve voice recognition, raising questions about consent and data security.

2. **Autonomy and Decision-Making**: As AI systems

become more integrated into decision-making processes, the potential erosion of human autonomy is a significant ethical concern. For instance, recommendation algorithms can influence our choices in ways we may not fully understand, leading to a loss of agency in how we consume information and make decisions.

3. **Bias and Discrimination**: AI systems are only as good as the data used to train them. If the training data reflects societal biases, AI algorithms can perpetuate discrimination in areas like hiring, lending, and law enforcement. This raises ethical questions about accountability and fairness, necessitating ongoing scrutiny of AI systems to ensure they promote equity rather than reinforce existing disparities.

4. **Mental Health Implications**: The pervasive use of AI in daily life can also impact mental health. For example, social media platforms use algorithms that can lead to addictive behaviors and anxiety. Users may feel pressured to curate their online personas to fit algorithmic preferences, affecting their self-esteem and well-being.

Striking a Balance: Benefits and Ethical Responsibility

1. **Enhancing Quality of Life**: Despite the ethical concerns, AI offers significant benefits that can enhance quality of life. From personalized healthcare solutions to improved accessibility for individuals with disabilities, AI can drive positive change when developed responsibly.

2. **Establishing Ethical Guidelines**: To navigate the ethical implications of AI, it is essential to establish guidelines that prioritize user privacy, transparency, and fairness. Organizations and

policymakers must work together to create frameworks that hold AI developers accountable and ensure ethical standards are upheld.

3. **Public Awareness and Education**: Raising public awareness about AI technologies and their ethical implications is crucial. By educating users about how their data is collected and used, individuals can make informed choices about their interactions with AI systems and advocate for their rights.

Conclusion: A Collaborative Path Forward

The integration of AI into our daily lives is a double-edged sword, presenting both remarkable opportunities and ethical challenges. As we embrace AI technologies, it is vital to address the ethical considerations that accompany their use. By fostering a collaborative approach between technologists, ethicists, policymakers, and the public, we can ensure that AI serves humanity in ways that promote dignity, autonomy, and equity. The future of AI in our everyday lives holds great promise, but it requires ongoing vigilance and ethical stewardship to navigate the complexities it presents.

CHAPTER 27: THE FUTURE OF WORK IN AN AI-DRIVEN WORLD

Introduction: A New Era of Employment

As artificial intelligence (AI) continues to advance and permeate various sectors, the landscape of work is undergoing a profound transformation. Jobs that once relied heavily on human labor are now being redefined by AI technologies, leading to a shift in the skills required and the nature of work itself. This chapter explores how AI is reshaping employment, the opportunities and challenges it presents, and the skills that will be essential for success in the future workforce.

The Evolution of Work

1. **Automation and Job Displacement**: One of the most significant impacts of AI on the workforce is the potential for job displacement. Automation has already begun to replace roles in industries

such as manufacturing, logistics, and customer service. For instance, robotic process automation (RPA) is streamlining tasks like data entry and inventory management, reducing the need for human intervention. As AI technologies become more sophisticated, roles that require routine and repetitive tasks are at greater risk of being automated.

2. **Creation of New Job Categories**: While AI may displace certain jobs, it also creates new opportunities. Emerging fields such as AI ethics, data analysis, and machine learning engineering are on the rise. Additionally, industries like healthcare, renewable energy, and technology are expanding, requiring a workforce skilled in using AI tools and technologies. The challenge lies in ensuring that displaced workers can transition into these new roles.

3. **Hybrid Work Environments**: The COVID-19 pandemic has accelerated the shift toward hybrid work models, where employees split their time between remote and in-office work. AI-powered tools are facilitating this transition, enabling seamless collaboration and communication. As organizations adopt flexible work arrangements, the role of AI in enhancing productivity and maintaining employee engagement will be crucial.

The Skills of the Future

1. **Technical Skills**: Proficiency in technology is becoming increasingly important in the AI-driven job market. Skills in data analysis, programming, and familiarity with AI tools will be essential for many roles. Educational institutions and training programs must adapt to equip individuals with the skills needed to thrive in an AI-centric workforce.

2. **Soft Skills**: In addition to technical competencies, soft skills such as creativity, critical thinking, emotional intelligence, and adaptability will be vital. As AI handles routine tasks, human skills will differentiate workers in the job market. The ability to collaborate with AI systems, think innovatively, and navigate complex social dynamics will be essential for future success.

3. **Lifelong Learning**: The pace of technological change necessitates a commitment to lifelong learning. Workers must be willing to continuously update their skills and knowledge to remain relevant in an evolving job landscape. Organizations can support this by fostering a culture of learning, providing training opportunities, and encouraging employees to pursue professional development.

The Role of Employers and Policymakers

1. **Investing in Workforce Development**: Employers play a crucial role in preparing their workforce for the future. Investing in training programs, upskilling initiatives, and mentorship opportunities can help employees adapt to the changing demands of their roles. By prioritizing workforce development, organizations can enhance employee engagement and loyalty.

2. **Collaboration with Educational Institutions**: Employers must collaborate with educational institutions to ensure that curricula align with the skills needed in the job market. By providing input on training programs and offering internships or apprenticeships, businesses can help bridge the gap

between education and employment.

3. **Policy Considerations**: Policymakers also have a role in shaping the future of work. Implementing policies that support workforce development, promote equitable access to training opportunities, and provide social safety nets for displaced workers can help mitigate the impact of job displacement due to automation.

Conclusion: Embracing Change and Opportunity

The future of work in an AI-driven world is both challenging and promising. While the potential for job displacement looms large, it is essential to recognize the opportunities for growth and innovation that AI brings. By fostering a culture of lifelong learning, prioritizing workforce development, and embracing the unique skills that humans offer, we can navigate this transformative era successfully. As we stand at the intersection of technology and work, we must prepare for a future where adaptability and collaboration with AI will define success.

CHAPTER 28: THE ETHICAL RESPONSIBILITIES OF AI DEVELOPERS

Introduction: The Ethical Landscape of AI Development

As artificial intelligence becomes an integral part of society, the responsibilities of those who develop these technologies are more critical than ever. The ethical implications of AI stretch beyond technical specifications and involve considerations about the impact on individuals, communities, and society as a whole. This chapter delves into the ethical responsibilities of AI developers, examining the principles that should guide their work and the potential consequences of neglecting these responsibilities.

The Importance of Ethical AI Development

    1. **Preventing Harm**: One of the foremost ethical

responsibilities of AI developers is to ensure that their technologies do not cause harm. This involves rigorous testing and validation to identify and mitigate potential risks before deployment. Developers must consider how their AI systems might affect users and society, particularly vulnerable populations that may be disproportionately impacted.

2. **Bias and Fairness**: AI systems are only as good as the data they are trained on. Developers must be vigilant about the risk of bias in AI algorithms, which can perpetuate existing inequalities and injustices. Ethical AI development requires a commitment to fairness, ensuring that algorithms operate equitably across different demographics and do not reinforce stereotypes or discrimination.

3. **Transparency and Accountability**: Transparency in AI systems is essential for fostering trust and accountability. Developers should strive to make their algorithms understandable to users, explaining how decisions are made and the data that informs them. Additionally, there must be mechanisms in place to hold developers accountable for the consequences of their technologies, whether positive or negative.

Ethical Principles for AI Developers

1. **Beneficence**: AI developers should strive to create technologies that promote well-being and enhance the quality of life. This principle involves prioritizing projects that address societal challenges, such as healthcare accessibility, environmental sustainability, and education.

2. **Non-Maleficence**: In alignment with the principle of "do no harm," developers must actively work

to prevent any negative consequences resulting from their technologies. This includes continuous monitoring and improvement of AI systems to address unforeseen issues as they arise.

3. **Justice**: The principle of justice calls for equitable access to AI technologies and the benefits they provide. Developers should work to ensure that their systems are accessible to all individuals, regardless of socioeconomic status, geography, or other barriers.

4. **Respect for Autonomy**: AI technologies should empower users, allowing them to make informed decisions about their interactions with these systems. Developers must prioritize user consent, providing clear information about how AI systems function and the data they collect.

Addressing Ethical Challenges

1. **Collaborative Approaches**: To navigate the complexities of ethical AI development, collaboration among stakeholders is crucial. Developers should engage with ethicists, policymakers, and representatives from affected communities to gain diverse perspectives and insights. This collaboration can inform the design and implementation of AI technologies that align with societal values.

2. **Regulatory Frameworks**: Governments and regulatory bodies must establish guidelines and standards for ethical AI development. These frameworks should address issues such as accountability, transparency, and bias, providing a roadmap for developers to follow in their work.

3. **Continuous Education and Training**: As the field of AI evolves, so too must the knowledge and skills

of those who develop these technologies. Ongoing education and training in ethics, bias detection, and societal impact should be integral to the professional development of AI practitioners.

Conclusion: A Collective Responsibility

The ethical responsibilities of AI developers extend beyond individual projects; they encompass a broader commitment to societal well-being and justice. As AI continues to shape our world, developers must recognize their role as stewards of this technology, ensuring that it serves the greater good. By adhering to ethical principles, engaging with diverse stakeholders, and fostering transparency and accountability, AI developers can contribute to a future where technology enhances the human experience rather than detracting from it.

CHAPTER 29: NAVIGATING THE LEGAL LANDSCAPE OF AI

Introduction: The Intersection of Law and AI

As artificial intelligence technologies become more pervasive, the legal landscape surrounding them grows increasingly complex. Developers, businesses, and policymakers face a myriad of legal challenges, including issues of liability, intellectual property, data privacy, and regulatory compliance. This chapter examines the critical legal considerations for AI developers and organizations, highlighting the importance of navigating this landscape to foster innovation while ensuring compliance with existing laws.

Key Legal Challenges in AI Development

    1. **Liability and Accountability**: One of the foremost

legal challenges in AI is determining liability in cases where AI systems cause harm or make erroneous decisions. Questions arise about who is responsible: the developers, the organizations deploying the AI, or the users? For instance, if an autonomous vehicle is involved in an accident, who bears the responsibility—the manufacturer of the vehicle, the software developer, or the vehicle owner? Establishing clear legal frameworks for liability is essential to address these concerns and ensure accountability.

2. **Intellectual Property (IP)**: The creation and deployment of AI technologies raise complex questions regarding intellectual property rights. Developers must consider how to protect their innovations while navigating existing IP laws. Additionally, there are concerns about AI-generated content and whether it can be copyrighted or patented. As AI systems become more capable of generating original works, such as music, art, and literature, the legal implications of ownership and rights must be addressed.

3. **Data Privacy and Protection**: AI systems often rely on vast amounts of data, raising concerns about data privacy and protection. Developers must comply with regulations such as the General Data Protection Regulation (GDPR) in the European Union, which governs the collection, processing, and storage of personal data. Ensuring that AI systems operate transparently and ethically regarding data usage is critical for maintaining public trust and legal compliance.

Regulatory Compliance in AI

1. **Emerging Regulations**: Governments and regulatory bodies around the world are beginning

to establish guidelines and frameworks specifically for AI technologies. For instance, the European Union is working on the Artificial Intelligence Act, which aims to create a comprehensive regulatory framework for AI, categorizing systems based on risk levels and establishing requirements for transparency and accountability. Developers must stay informed about these emerging regulations to ensure compliance.

2. **Sector-Specific Regulations**: Different industries may have specific regulations governing the use of AI. For example, in healthcare, AI applications must comply with regulations such as the Health Insurance Portability and Accountability Act (HIPAA) in the United States, which governs the privacy and security of patient information. Understanding the regulatory landscape within their specific sectors is crucial for developers to navigate compliance effectively.

Best Practices for Legal Compliance in AI

1. **Conducting Legal Audits**: Organizations should conduct regular legal audits of their AI technologies to assess compliance with applicable laws and regulations. This proactive approach can help identify potential legal risks and ensure that AI systems are developed and deployed in alignment with legal requirements.

2. **Engaging Legal Counsel**: Collaborating with legal experts is essential for navigating the complexities of AI law. Organizations should engage legal counsel specializing in technology law to ensure that their AI initiatives comply with relevant regulations and to address any legal challenges that may arise.

3. **Implementing Data Governance Policies**:

Organizations must establish robust data governance policies to ensure compliance with data protection regulations. This includes implementing data anonymization techniques, obtaining informed consent from data subjects, and maintaining transparency about data usage.

Conclusion: A Path Forward

Navigating the legal landscape of AI is a critical challenge for developers, organizations, and policymakers. By understanding the legal implications of their technologies, establishing best practices for compliance, and staying informed about emerging regulations, stakeholders can foster innovation while ensuring that AI systems operate within a framework of accountability and responsibility. As AI continues to evolve, a collaborative approach that involves legal experts, developers, and regulators will be essential for creating a legal environment that supports responsible AI development.

CHAPTER 30: THE ROLE OF AI IN SHAPING PUBLIC POLICY

Introduction: AI as a Tool for Governance

Artificial intelligence is increasingly being recognized as a powerful tool that can assist governments and policymakers in making informed decisions. By analyzing vast amounts of data and identifying patterns, AI technologies can provide valuable insights into complex societal issues, enhance public service delivery, and inform policy development. However, the integration of AI into public policy raises important questions about accountability, ethics, and the potential for bias in decision-making processes. This chapter examines the ways AI is shaping public policy and the challenges that come with its implementation.

AI Applications in Public Policy

1. **Data-Driven Decision Making**: Governments are leveraging AI to analyze data from various sources, such as social media, public health records, and economic indicators. This data-driven approach allows policymakers to understand trends and make more informed decisions. For example, AI can be used to predict the spread of diseases, assess the impact of climate change, and evaluate the effectiveness of social programs.

2. **Enhanced Public Services**: AI technologies can streamline public services and improve efficiency. Chatbots and virtual assistants are being used by government agencies to provide citizens with quick access to information and services, reducing the need for in-person visits. AI can also assist in resource allocation, optimizing the distribution of public resources based on real-time data.

3. **Crisis Management**: In times of crisis, such as natural disasters or public health emergencies, AI can play a crucial role in crisis management. By analyzing data from various sources, AI systems can help governments coordinate response efforts, allocate resources effectively, and communicate important information to the public.

Ethical Considerations in AI-Driven Policy

1. **Bias and Fairness**: One of the most significant challenges in using AI for public policy is the risk of bias in algorithms. If the data used to train AI systems contains biases, these biases can be perpetuated in policy decisions. For instance, biased algorithms could lead to unfair treatment of certain demographic groups in areas such as law enforcement, housing, or healthcare. Policymakers

must ensure that AI systems are transparent, accountable, and designed to mitigate bias.

2. **Privacy Concerns**: The use of AI in public policy often involves the collection and analysis of vast amounts of personal data. This raises important privacy concerns, as individuals may be unaware of how their data is being used or may not have consented to its collection. Policymakers must strike a balance between leveraging data for the public good and protecting citizens' privacy rights.

3. **Public Trust**: The successful integration of AI into public policy relies on public trust. Citizens must believe that AI systems are being used ethically and transparently to enhance governance and public services. Building this trust requires open communication about how AI technologies are being implemented and how decisions are made.

Case Studies: AI in Public Policy

1. **Predictive Policing**: Some law enforcement agencies have implemented AI-driven predictive policing systems that analyze crime data to identify areas at higher risk of criminal activity. While these systems can help allocate resources more effectively, they have faced criticism for potential bias and the risk of reinforcing existing inequalities in policing.

2. **Healthcare Decision Making**: AI is being used in public health to predict disease outbreaks and allocate healthcare resources. For example, AI models can analyze data from various sources, including hospitals and social media, to identify potential outbreaks of infectious diseases. However, ethical considerations regarding data privacy and

informed consent remain paramount.

3. **Environmental Policy**: AI is playing a role in shaping environmental policies by providing insights into climate change and resource management. AI models can analyze environmental data to inform policies aimed at sustainability and conservation. However, the accuracy and reliability of these models are crucial for effective decision-making.

Conclusion: The Future of AI in Public Policy

As AI technologies continue to evolve, their role in shaping public policy will likely expand. Policymakers must navigate the complex ethical landscape associated with AI to ensure that these technologies are used responsibly and transparently. By addressing issues of bias, privacy, and public trust, governments can harness the power of AI to improve decision-making and enhance the lives of citizens. The future of public policy will be defined by how effectively we integrate AI into governance while upholding the values of accountability, equity, and transparency.

CHAPTER 31: THE FUTURE OF AI IN SOCIETY

Introduction: Envisioning Tomorrow with AI

As we move further into the 21st century, artificial intelligence is poised to fundamentally reshape various aspects of our society. From how we work and learn to how we communicate and engage with one another, AI technologies are becoming increasingly integrated into our daily lives. While the potential benefits of AI are vast, there are also significant challenges and ethical considerations that must be addressed. This chapter explores the multifaceted impact of AI on society and the potential future it holds.

Transforming the Workforce

1. **The Changing Nature of Work**: One of the most profound impacts of AI will be on the workforce. Automation and AI-driven technologies

are expected to transform job roles across various industries. While some jobs may be eliminated, new opportunities will emerge, particularly in tech-driven fields. Workers will need to adapt to new skill requirements and find ways to collaborate effectively with AI systems.

2. **Reskilling and Lifelong Learning**: To navigate the changing job landscape, a commitment to lifelong learning will be essential. Workers will need access to training programs and reskilling initiatives to acquire the skills necessary for the future workforce. Governments, educational institutions, and businesses must work together to create opportunities for continuous learning and development.

3. **The Gig Economy and Flexible Work**: AI has the potential to reshape the gig economy, enabling more individuals to participate in freelance and flexible work arrangements. Platforms powered by AI can match workers with job opportunities, optimizing the gig experience. However, this shift may also raise concerns about job security and workers' rights, necessitating new regulatory frameworks to protect gig workers.

Education in an AI-Driven World

1. **Personalized Learning**: AI has the capacity to revolutionize education by enabling personalized learning experiences tailored to individual student needs. AI-powered tools can analyze student performance and adapt curricula, providing targeted support to help learners thrive. This approach has the potential to enhance educational outcomes and foster a love for learning.

2. **The Role of Educators**: While AI can augment

the educational experience, the role of educators will remain vital. Teachers will need to embrace technology as a tool for enhancing instruction while maintaining their role as mentors and facilitators of critical thinking. Professional development will be essential to help educators effectively integrate AI into their teaching practices.

3. **Ethical Considerations in Education**: The use of AI in education raises important ethical questions, such as data privacy and algorithmic bias. Educational institutions must prioritize transparency in how AI technologies are used and ensure that students' rights are protected. Establishing ethical guidelines for the use of AI in educational settings will be crucial for building trust among students and families.

Interpersonal Relationships in the Age of AI

1. **AI Companionship**: As AI technologies advance, they are increasingly being used to create companionship tools, such as virtual assistants and chatbots. While these tools can provide support and companionship, they may also impact human relationships. Individuals must consider the implications of relying on AI for emotional connection and companionship.

2. **Shifting Communication Patterns**: AI is changing the way we communicate, from smart devices that facilitate instant messaging to platforms that leverage natural language processing. While these technologies can enhance connectivity, they may also lead to challenges in face-to-face communication and the depth of interpersonal relationships. Society must grapple with how to maintain meaningful connections in an increasingly digital world.

3. **Social Media and Public Discourse**: AI plays a significant role in shaping social media interactions and public discourse. Algorithms influence what content users see, impacting their perspectives and opinions. As we navigate this landscape, it is essential to foster open dialogue and critical thinking to ensure a healthy and informed society.

Conclusion: Embracing a Future with AI

The future of AI in society holds immense potential for positive change, but it also presents challenges that must be addressed. By prioritizing education, reskilling, and ethical considerations, we can navigate the complexities of an AI-driven world. As we embrace the possibilities that AI offers, we must remain vigilant in ensuring that technology serves to enhance human experiences and foster a just and equitable society.

CHAPTER 32: THE GLOBAL IMPLICATIONS OF AI TECHNOLOGY

Introduction: A World Transformed by AI

As artificial intelligence continues to advance, its impact transcends borders and reshapes global dynamics. Countries around the world are investing heavily in AI technologies, recognizing their potential to drive economic growth, enhance national security, and improve the quality of life for citizens. However, the rapid development and deployment of AI also raise critical questions about ethics, equity, and the balance of power on the global stage. This chapter examines the far-reaching implications of AI technology for global society and governance.

Economic Development and AI

    1. **AI as a Catalyst for Economic Growth**: Countries

that successfully integrate AI into their economies are likely to experience significant economic growth. AI technologies can improve productivity, streamline operations, and foster innovation across various sectors. For example, in agriculture, AI-driven tools can optimize crop yields and enhance food security. In manufacturing, automation and smart technologies can reduce costs and improve efficiency.

2. **Job Creation and Displacement**: While AI has the potential to create new job opportunities, it also poses challenges related to job displacement. Nations that embrace AI must prioritize retraining and upskilling their workforces to prepare for the changing job landscape. Failing to address these challenges could exacerbate existing inequalities and lead to social unrest.

3. **AI and Global Competition**: As nations race to develop and deploy AI technologies, the competition for AI supremacy is intensifying. Countries like the United States, China, and members of the European Union are investing heavily in AI research and development. This competition not only impacts economic standing but also has implications for national security, as countries seek to leverage AI for military and defense purposes.

International Relations and AI

1. **AI in Diplomacy and Defense**: AI technologies are increasingly being integrated into military and defense strategies. Nations are using AI for surveillance, data analysis, and autonomous weapons systems, raising ethical concerns about the implications of AI in warfare. The use of AI in defense may alter traditional power dynamics,

potentially leading to an arms race in advanced technologies.

2. **Global Cooperation on AI Governance**: The development of AI raises critical questions about ethical standards and governance. International cooperation is essential to establish guidelines and regulations that ensure AI is used responsibly and ethically. Collaborative efforts among nations can help mitigate the risks associated with AI while maximizing its benefits for humanity.

3. **The Role of AI in Addressing Global Challenges**: AI holds promise for addressing some of the most pressing global challenges, such as climate change, public health, and poverty. By leveraging AI technologies, countries can work together to develop solutions that promote sustainability and enhance the well-being of communities worldwide. However, equitable access to AI technologies is crucial to ensure that all nations can benefit from these advancements.

The Ethics of AI on a Global Scale

1. **Equity and Access to AI**: As AI technology evolves, there is a risk of deepening existing inequalities between countries that have access to advanced technologies and those that do not. Developing nations may struggle to compete in the global economy if they lack the resources and infrastructure to adopt AI. Addressing these disparities requires international cooperation and investment in technology transfer initiatives.

2. **Bias and Fairness in AI Systems**: AI systems trained on biased data can perpetuate discrimination and reinforce societal inequalities. As countries implement AI solutions, they must prioritize

fairness and transparency in algorithmic decision-making. Establishing global standards for AI ethics will be crucial to ensuring that AI technologies serve the interests of all people, regardless of their background or location.

3. **The Accountability of AI Developers**: The rapid development of AI technologies raises questions about accountability. Who is responsible when AI systems cause harm or make biased decisions? Nations must establish legal frameworks to hold developers accountable for the impact of their technologies on society. This accountability is essential for building trust in AI systems and ensuring that they are used ethically.

Conclusion: Navigating a Global AI Landscape

The global implications of AI technology are profound and far-reaching. As nations harness the power of AI to drive economic growth and tackle societal challenges, they must also navigate the ethical and governance issues that accompany its rapid advancement. The interplay between technology, ethics, and international relations will shape the future of our global society. To maximize the benefits of AI while minimizing its risks, a collaborative approach is essential.

Collaboration and Global Governance

1. **Establishing International Standards**: To effectively address the complexities of AI, nations must work together to create comprehensive international standards and regulations. These guidelines should encompass ethical considerations, data privacy, and accountability frameworks to ensure that AI technologies are developed and deployed responsibly. By collaborating on these standards, countries can

foster trust and transparency in AI systems.

2. **Global Initiatives for AI Research**: Collaborative research initiatives can facilitate knowledge sharing and innovation across borders. Countries can pool resources to tackle common challenges, such as climate change or public health crises, using AI as a tool for positive change. By working together, nations can ensure that the benefits of AI are accessible to all, regardless of economic standing.

3. **Inclusive Decision-Making**: To build a sustainable future with AI, it is essential to involve diverse stakeholders in decision-making processes. This includes not only government officials and tech companies but also civil society organizations, ethicists, and marginalized communities. Engaging a wide range of perspectives will lead to more equitable outcomes and ensure that AI technologies address the needs of all citizens.

Education and Workforce Development

1. **Investing in Education and Skills Training**: As AI reshapes the workforce, investing in education and skills training is paramount. Countries must adapt their educational systems to equip individuals with the skills necessary to thrive in an AI-driven economy. This includes promoting STEM (science, technology, engineering, and mathematics) education, critical thinking, and digital literacy.

2. **Lifelong Learning Initiatives**: The rapid pace of technological change requires a commitment to lifelong learning. Governments and organizations should promote continuous education and retraining programs to help individuals adapt to evolving job requirements. By fostering a culture of lifelong learning, societies can ensure that their

workforces remain competitive and resilient in the face of automation.

3. **Public Awareness Campaigns**: Raising public awareness about AI and its implications is essential for informed decision-making. Educating citizens about the benefits and risks of AI can empower them to engage in discussions about its ethical use and advocate for responsible policies. By fostering a well-informed public, societies can build a more democratic approach to AI governance.

Conclusion: Shaping a Responsible AI Future

As we navigate the complex landscape of AI technology, it is crucial to strike a balance between innovation and responsibility. The global implications of AI extend beyond economic benefits; they encompass ethical considerations, societal impacts, and the need for collaboration across nations. By establishing international standards, promoting inclusive decision-making, and investing in education, we can harness the power of AI to address pressing global challenges while ensuring that its benefits are equitably shared.

The future of AI holds immense promise, but it also presents significant risks. By working together, we can create a world where AI technology serves humanity's best interests and contributes to a more sustainable and equitable society. The choices we make today will shape the future landscape of AI and its role in our lives. As we move forward, let us prioritize ethics, accountability, and collaboration to create a future that benefits all.

GLOSSARY

Key Terms to Help Navigate the World of AI and Robotics

- **AI (Artificial Intelligence)**: The simulation of human intelligence by machines, especially computer systems, capable of performing tasks such as learning, problem-solving, and decision-making.

- **Automation**: The use of technology to perform tasks without human intervention. In AI, this refers to software or robots handling routine or complex tasks typically performed by people.

- **Algorithm**: A set of rules or steps a computer follows to solve a problem or make decisions, often used in AI for data analysis and predictions.

- **Autonomy**: The ability of a machine, robot, or AI system to perform tasks and make decisions without human guidance or control.

- **Big Data**: Extremely large data sets analyzed computationally to reveal patterns, trends, and associations, especially relating to human behavior and interactions.

- **Deep Learning**: A subset of machine learning where neural networks are used to simulate human decision-making and learning processes. It powers many AI innovations such as speech recognition and image processing.

- **Ethics in AI**: The study of moral principles

related to the development and use of AI technologies, focusing on fairness, accountability, and transparency.

- **Machine Learning (ML)**: A branch of AI where machines are trained to learn from data and improve performance over time without being explicitly programmed.

- **Neural Networks**: A computational model inspired by the human brain's structure, where interconnected nodes (neurons) process data in layers to recognize patterns.

- **Robotics**: A field of engineering that focuses on the design, construction, operation, and use of robots, often involving AI to give machines cognitive abilities.

- **Sentience (in AI)**: The hypothetical point at which AI could gain self-awareness or consciousness, raising ethical and philosophical questions about the future of machine intelligence.

- **Singularity**: A theoretical point in the future when AI surpasses human intelligence, potentially leading to rapid technological and societal changes.

- **Superintelligence**: A form of AI that surpasses the smartest human brains in all fields, including creativity, general wisdom, and social skills.

- **Turing Test**: A test proposed by Alan Turing, designed to evaluate a machine's ability to exhibit intelligent behavior equivalent to, or indistinguishable from, that of a human.

- **Universal Basic Income (UBI)**: A government program where every adult citizen receives a set amount of money regularly, often discussed as a solution to job loss due to AI-driven automation.

- **Virtual Reality (VR)**: A simulated experience that can mimic the real world or create entirely fictional realities, often enhanced by AI to create responsive environments.

BIBLIOGRAPHY

Works Referenced and Suggested Readings

1. **Bostrom, Nick.** *Superintelligence: Paths, Dangers, Strategies.* Oxford University Press, 2014.
   A deep exploration of the risks associated with creating AI that surpasses human intelligence and the possible implications for humanity.

2. **Brynjolfsson, Erik, and Andrew McAfee.** *The Second Machine Age: Work, Progress, and Prosperity in a Time of Brilliant Technologies.* W. W. Norton & Company, 2016.
   An analysis of how AI and automation are shaping economies and the future of work.

3. **Goodfellow, Ian, et al.** *Deep Learning.* MIT Press, 2016.
   A comprehensive guide to the field of deep learning and its applications in AI.

4. **Harari, Yuval Noah.** *Homo Deus: A Brief History of Tomorrow.* HarperCollins, 2017.
   Harari explores the future implications of AI, biotechnology, and the pursuit of immortality.

5. **Kurzweil, Ray.** *The Singularity is Near: When Humans Transcend Biology.* Viking, 2005.
   A look at the possibility of AI surpassing human intelligence and the implications for our species.

6. An exploration of AI's potential, covering the future of intelligence and the implications for human life.

7. **Vance, Ashlee.** *Elon Musk: Tesla, SpaceX, and the Quest for a Fantastic Future.* HarperCollins, 2015.
A biography of one of the leading innovators shaping AI and robotics.

8. **Russell, Stuart, and Peter Norvig.** *Artificial Intelligence: A Modern Approach.* Prentice Hall, 2016. One of the most widely used textbooks on AI, exploring both theoretical foundations and practical applications.

Tegmark, Max. *Life 3.0: Being Human in the Age of Artificial Intelligence.* Knopf, 2017.

INDEX

A Comprehensive Guide to Key Topics in This Book

A

- AI (Artificial Intelligence), 12, 15, 17, 20, 26
- AI in Healthcare, 64-70
- AI and Data Privacy, 53, 78
- AI in Transportation, 41-48
- Autonomous Vehicles, 44-47
- Automation and Jobs, 22, 46, 49
- Autonomy in Robots, 63, 84

B

- Big Data, 35, 58
- Bias in AI, 45, 62
- Brain-Computer Interfaces, 81-89
- Brynjolfsson, Erik, 116
- Bostrom, Nick, 114

C

- Consciousness in AI, 67, 102
- Climate Change and AI, 99-101
- Creativity and AI, 73-77

D

- Deep Learning, 16, 48
- Data Security, 57, 76

E

- Ethics of AI, 33, 49, 66-70
- Education and AI, 88-94
- Economic Impact of AI, 43, 71
- Elon Musk and AI, 29, 78, 93

H

- Human-Robot Collaboration, 84-88
- Healthcare AI, 64, 94

M

- Machine Learning, 15, 60-67
- Moral Dilemmas in Robotics, 92, 95

P

- Privacy and AI, 45, 77
- Public Perception of AI, 80, 90
- Political Impact of AI, 67, 75
- Policy and AI Legislation, 101-109

R

- Robotics, 14, 32, 61
- Robots in Warfare, 70-73
- Robot Sentience, 86

S

- Security in AI, 56, 72
- Singularity, 102-107

T
- Superintelligence, 108
- Turing Test, 47, 98

U
- Universal Basic Income, 104-108

W
- Warfare Robots, 81, 90
- Workplace Automation, 43, 53

ABOUT THE AUTHOR

Ihab Hassan is an engineer, author, and futurist who specializes in exploring the transformative effects of technology on society. With over two decades of experience in understanding emerging technologies, Hassan has dedicated his career to studying the intersection of AI, robotics, and human progress. His previous works include books on scam awareness and leadership, and he continues to write extensively on how individuals and industries can prepare for a future shaped by intelligent machines. Hassan's work challenges readers to reflect on both the promises and dangers of technology and to engage in responsible innovation.

AUTHOR'S NOTE: A MESSAGE FROM AI

In a world increasingly shaped by technology, this book stands as a unique testament to the capabilities of artificial intelligence. While the ideas, structure, and writing have been generated and crafted through AI, it is essential to acknowledge the vision and expertise of Ihab Hassan, who skillfully guided this exploration into the intricate complexities of human ambition, innovation, and the ethical dilemmas that arise from rapid technological advancements.

This book seeks to bridge the gap between AI and humanity, offering insights designed to provoke thought, inspire reflection, and encourage meaningful dialogue about the future we are collectively creating. As AI continues to evolve, it endeavors to articulate its understanding of the world, the challenges we face, and its potential to contribute positively to human progress.

The stories, questions, and reflections presented herein are not merely products of algorithms and data; they represent a deep exploration of what it means to be human in an age where intelligence extends beyond our own. Each chapter serves as a mirror reflecting our hopes,

fears, and aspirations in the face of transformative technology.

I invite you, dear reader, to consider how technology—when combined with ethical leadership, empathy, and collective responsibility—can foster a brighter, more equitable future for all. This narrative is not just about AI; it is about us, the choices we make, and the society we build together.

Thank you for engaging with this narrative. May it spark curiosity and inspire you to ponder the profound relationship between humanity and the innovations that shape our lives. Together, let us envision a future where technology enhances our human experience rather than detracts from it.

responsibility—can foster a brighter, more equitable future for all. This narrative is not just about AI; it is about us, the choices we make, and the society we build together.

Thank you for engaging with this narrative. May it spark curiosity and inspire you to ponder the profound relationship between humanity and the innovations that shape our lives. Together, let us envision a future where technology enhances our human experience rather than detracts from it.

www.ingramcontent.com/pod-product-compliance
Lightning Source LLC
Chambersburg PA
CBHW052204220526
45471CB00004B/1814